Lecture Notes in Computer Science 721
Edited by G. Goos and J. Hartmanis

John Fitch (Ed.)

Design and Implementation of Symbolic Computation Systems

International Symposium, DISCO '92
Bath, U.K., April 13-15, 1992
Proceedings

Springer-Verlag
Berlin Heidelberg New York
London Paris Tokyo
Hong Kong Barcelona
Budapest

Series Editors

Gerhard Goos
Universität Karlsruhe
Postfach 69 80
Vincenz-Priessnitz-Straße 1
D-76131 Karlsruhe, Germany

Juris Hartmanis
Cornell University
Department of Computer Science
4130 Upson Hall
Ithaca, NY 14853, USA

Volume Editor

John Fitch
School of Mathematical Sciences
University of Bath, Bath, England BA2 7AY

CR Subject Classification (1991): D.1, D.2.1, D.2.10, D.3, I.1, I.2.2-3, I.2.5, I.3.5-6

ISBN 3-540-57272-4 Springer-Verlag Berlin Heidelberg New York
ISBN 0-387-57272-4 Springer-Verlag New York Berlin Heidelberg

Printed in Germany

Typesetting: Camera-ready by author
Printing and binding: Druckhaus Beltz, Hemsbach/Bergstr.
45/3140-543210 - Printed on acid-free paper

Preface

DISCO '92 was held from April 13 to 15, 1992, in the ancient city of Bath, England. The actual site was the Newton Park campus of Bath College of Higher Education. Besides the formal lectures dedicated to Design and Implementation issues of Computer Algebra, there were several software demonstrations, and the opportunity for system designers to compare systems. The demonstrations were facilitated by the loan of equipment from the University of Bath and from Codemist Ltd.

The schedule allowed time for discussions after the papers. One general theme which clearly emerged was the need for interconnections between systems, as no one system will have in it all the facilities that the users want. There are the various efforts going on to design such links, but generally in limited contexts (such as the Maple project or the Posso project).

Many people helped in the organisation and running of DISCO '92, but I would like to make a special mention of John Fitch, who did much of the work when, at the last minute, I had other administrative tasks to perform.

July 1993 J. H. Davenport

Programme Committee

J. H. Davenport (Chairman)
A. Bundy
C. M. Hoffmann
A. Kreczmar
A. Miola
P. S. Wang

Local Organisation
J. P. Fitch

Table of Contents

Template-based Formula Editing in Kaava

Ken Rimey*

rimey@cs.hut.fi
Department of Computer Science
Helsinki University of Technology

Abstract. This paper describes a user interface for entering mathematical formulas directly in two-dimensional notation. This interface is part of a small, experimental computer algebra system called Kaava. It demonstrates a data-driven structure editing style that partially accommodates the user's view of formulas as two-dimensional arrangements of symbols on the page.

1 Introduction

Before someone with a mathematical problem to solve can benefit from having a workstation on his desk, he has to enter the relevant formulas onto the screen. All the major computer algebra systems have expected him to use a Fortran-like expression syntax for this. Although we have grown accustomed to using this syntax for the kinds of expressions we put into Fortran programs, the richer, more concise, and easier-to-read notation that a scientist or applied mathematician uses in problem solving does not take the form of strings of characters. Although the major computer algebra systems use character strings for input, all support two-dimensional mathematical notation for output. It would be better to use the same language for both input and output.

This has been done, both in small algebra systems and in add-on interfaces to major algebra systems. Neil Soiffer's PhD thesis surveys these efforts and provides a comprehensive discussion of strategies for directly entering two-dimensional notation [8]. The key issue seems to be a conflict between the hierarchical structure of mathematical expressions and our tendency to read and write from left to right.

Consider the expression

$$ax^2 + bx + c$$

In reading it aloud, or in writing it, most people would begin with the a. Even the purest mathematician is aware of the left-to-right arrangement of the symbols on the page. The natural basis for representing the expression in a computer, on the other hand, is its hierarchical structure as a sum of three terms, two of which are products, and so on. Designing an editor that represents expressions as trees and yet allows the user his visual bias in entering them is a hard problem.

Soiffer favors retaining an underlying string model as the basis for the editing operations:

```
a * x ^ 2 + b * x + c
```

* Author's current address: Unda Oy, Ahventie 4A, SF-02170 Espoo, Finland.

This string reflects the left-to-right layout of the displayed expression and is second nature for a programmer to type. Soiffer discusses two alternative ways of keeping the expression tree up to date as the user, in effect, edits a string: incremental parsing and the simulation of parsing through tree manipulations. The latter was first considered by Kaiser and Kant [4].

This paper focuses on the contrasting approach that completely abandons Fortran-like expression languages. The small, Macintosh-based algebra system, *Milo*, has demonstrated how this can be done without the clumsiness generally associated with structure editing [5]. One of the key ideas is allowing the selection of an insertion point. This seemingly minor user interface issue is trickier and more important than it may at first seem. Milo has recently been enhanced and incorporated into the *FrameMaker* desk-top publishing system [2, 3].

The experimental small algebra system, *Kaava*,[2] takes a similar approach. In designing its formula editor, I have endeavored to create a data-driven mechanism that can be exposed to the user. The remainder of this paper describes the result. After some comments on the representation of the expression trees and on the user extensibility, I describe the editing strategy in three stages with progressively fewer restrictions on what can be selected: placeholders only, arbitrary expressions, and finally arbitrary expressions and insertion points.

Kaava is written in Common Lisp and runs under the X window system. Harri Pasanen provides an overview of the system at an early stage of development in his master's thesis [6].

2 Representation

Kaava represents each formula displayed on the screen as an expression tree. In fact, it represents the user's entire working document (containing formulas and paragraphs of explanatory text) as one big tree. Each node of the tree contains the following information:

- An operator. Leaf nodes have operators like *number* and *variable*.
- A list of child nodes, i.e. arguments.
- Additional information, depending on the operator (number's value, variable's name, etc.).

Although the objects that represent these abstract nodes in the program have other information attached to them, including pointers to a data structure describing the layout of the symbols on the screen, this is unimportant in understanding the program's behavior. The user edits an expression tree.

The task of choosing a vocabulary of operators is complicated by the conflicting needs of editing and algebraic manipulation. Instead of compromising, Kaava uses different representations for the two purposes, converting expressions freely back and forth. The editing representation is the primary one, and the only one of concern in this paper. The other is used in performing mathematical commands.

Whereas a small operator vocabulary is most conducive to algebraic manipulation, Kaava's editing representation is designed to correspond closely to what the

[2] *Kaava* rhymes with *java* and is the Finnish for *formula*.

user sees on the screen. For example, rather than abandon the division operator in favor of negative exponents, Kaava instead represents the expression

$$\frac{xy}{2ab}$$

straightforwardly as a quotient of two products. Rather than abandon unary minus in favor of multiplication by -1, Kaava uses the unary minus operator wherever a minus sign appears in an expression, as in $-x$, $-2x$, and even -3. A subtraction operator, on the other hand, would be undesirable because of the unwanted structure it would impose on an expression such as the following:

$$a + b - c - d$$

Kaava represents this as a sum of four terms, and by the way, has no preference as to their order.

Figure 1 illustrates these conventions using the tree for the expression

$$\frac{x^2 - 2x + 1}{(x+1)(x-1)}$$

It also illustrates our use of an explicit parenthesizing operator. After each editing

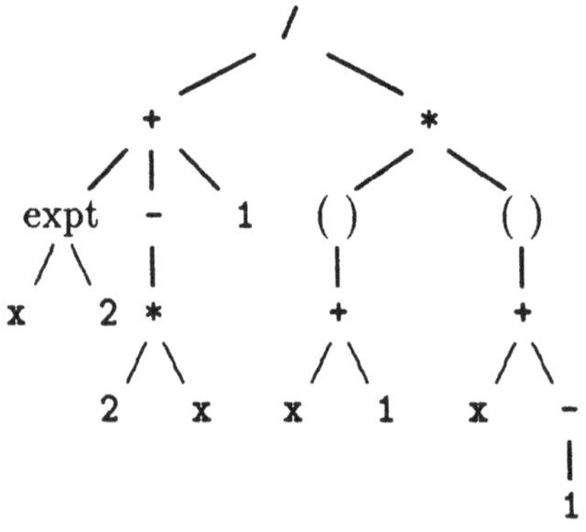

Fig. 1. An expression tree.

operation, Kaava inserts instances of this operator wherever it is needed to make the expression tree legal. It also *flattens* nested instances of certain operators.

Legality is determined by a *structural schema*. This defines how many arguments an operator can have and what the operators of these arguments can be.[3] For example, a child of a product node may not be a + node, but it may be a parenthesizing node.

[3] The general idea of a structural schema appears to be well-known in the structure editing community. The variant used in Kaava is relatively weak insofar as it does not have the power of a context-free tree grammar.

We use a combination of mechanisms to specify the schema. First, we assign each operator to one or more operator classes and specify the classes to which its arguments may belong. Second, we give each operator a precedence value and specify minimum precedence values for its children. The operator-class mechanism is useful for enforcing basic sanity rules. For instance, it is important to insure that the argument of a product is never a paragraph of text. If, as in this case, flattening and adding parentheses does not make the final tree legal, the program refuses to perform the editing operation.

The purpose of flattening is to facilitate the entry of operators that take arbitrary numbers of arguments. Without it, substituting $c + d$ for the c in

$$a + b + c$$

would produce

$$a + b + (c + d)$$

The actual result is

$$a + b + c + d$$

Kaava flattens not just associative mathematical operators, but certain nonmathematical operators as well. For example, there is a *comma* operator that takes two or more arguments and appears, for instance, in the representation of

$$f(a, b, c)$$

A command to replace the c by "c, d" results in the expression

$$f(a, b, c, d)$$

3 Pasting

In a sense, there is only one important command in Kaava: *pasting.* It can be invoked either with the mouse or from the keyboard. The pasted expression determines the semantics of the operation.

In the simplest case, pasting replaces the *selection* by the expression being pasted. Consider again the example of the sum $a + b + c$. Selecting the c

$$a + b + \boxed{c}$$

and pasting $c + d$ results, after flattening, in the expression

$$a + b + c + d$$

The initial selection can be done with the left mouse button. The pasting can then be done by selecting any $c + d$ in the document with the right mouse button. The left button controls the *primary selection* (the destination), while the right button controls the *copy selection* (the source). Only the primary selection is persistent; the copy selection disappears when the mouse button is released.

We call pasting done with the mouse in this way *copy-pasting*. It combines two operations from the standard cut-copy-paste trio, eliminating the cut buffer. Although the user interface is spartan, the effect of copy-paste is the same as that of an intra-application drag-and-drop.

The user can paste an expression from the keyboard by hitting a key bound to the expression. Keys are bound to expressions by including *keybindings* in the document. For example, the key Control-A can be bound to the expression $c + d$ by putting the keybinding

$$\texttt{C} - \texttt{A} : c + d$$

anywhere in the document. Then the pasting operation of the previous example can be performed by hitting that key. The default initial document predefines many keybindings, including, for instance, bindings of the alphabetic keys to their corresponding symbols:

$$\texttt{a} : a \;\; \texttt{b} : b \;\; \texttt{c} : c \ldots$$

The simple replacement semantics generalize to the semantics of pasting a *template*. The initial document includes, for example, the following:

$$+ : .?{+}?. \quad * : .??. \quad / : \frac{.?}{?.}$$

Template pasting semantics is the topic of sections 4–6.

Unfortunately, the behavior that users expect of the "-" key is too complicated to be specified in this way. Sometimes it should paste one template ($-?.$), and sometimes another ($.?-?.$). It actually invokes a built-in command, represented in the document by its quoted name:

$$- : \text{“minus”}$$

Sometimes the effect of pasting a built-in command may have nothing to do with pasting templates:

$$\texttt{C} - \texttt{Q} : \text{“quit”}$$

Kaava's as-yet limited mathematical facilities are invoked through the pasting of rewrite rules onto expressions to be transformed. Here it makes sense to have many bindings for a single key, as in the expand command:

$$\texttt{C} - \texttt{X} : \boldsymbol{a(x + y) = ax + ay}$$
$$\texttt{C} - \texttt{X} : \boldsymbol{(x + y)^n} = \text{expand-expt}\left(\boldsymbol{(x + y)^n}\right)$$

Most everything that can be done with keybindings can be done as well with copy-pasting. Collecting the standard keybindings neatly at the head of the document produces a kind of palette from which the user can copy-paste as an alternative to the keyboard.

The special behavior of rules and built-in commands makes them tricky to manipulate. Binding the middle mouse button to a *literal copy-paste* operation with only simple replacement semantics solves this problem.

4 Placeholders

Placeholders enable the representation of incomplete expressions. They appear in expressions that are in the process of being entered:

$$ax^2{+}?$$

They also appear in templates (expressions meant for pasting):

$$+ :?{+}?$$

Kaava displays ordinary placeholders as question marks.

If we select only placeholders, the pasting of a template is purely a matter of replacement:

$$ax^2 + \boxed{?} \xrightarrow{?+?} ax^2{+}?{+}?$$

Here the template is ?+?.

Arbitrarily complex expressions can be entered in this way. For example, the expression

$$ax + b$$

can be entered as follows:

$$\boxed{?} \xrightarrow{?+?} \boxed{?}{+}? \xrightarrow{??} \boxed{?}?{+}? \xrightarrow{a} a\boxed{?}{+}? \xrightarrow{x} ax + \boxed{?} \xrightarrow{b} ax + b$$

With appropriate keybindings, what the user types is

```
+ * a x b
```

Unfortunately, successive placeholder replacement amounts to using prefix syntax for input.

Another problem is that the user apparently needs to select a placeholder before each keystroke. In practice, having to switch back and forth between the mouse and keyboard in entering an expression is unacceptable; the system must always leave something selected after each operation. Kaava does this, but it does not always select a placeholder. Whenever there are placeholders in the replacement expression, however, it automatically selects one of them.

5 Selections

The opportunity to go beyond replacement semantics makes the selection of arbitrary expressions attractive. The user establishes a selection by depressing the left mouse button, moving the mouse, and then releasing the button. The selected expression is the smallest one whose bounding box includes both endpoints.[4]

The user can adjust the selection with various built-in commands. With the standard keybindings, the four arrow keys move the selection up, down, left, and right in the expression tree: Up-arrow selects the parent; Down-arrow, the first child;

[4] Using a simple bounding box would not be appropriate if Kaava broke long expressions onto multiple lines.

Left-arrow and Right-arrow, a sibling. The space bar expands the selection like Up-arrow. The tab key moves it to a nearby placeholder. (These keybinding conventions come from Milo.)

When a template containing at least one placeholder is pasted onto a target expression, the replacement expression is constructed by incorporating the target into the template in place of one of the placeholders. The authors of CaminoReal call this *wrapping* [1]. Soiffer would call the template an *overlay* [8]. We have so far been satisfied with the choice of the first placeholder in the template as the one to replace, but it would be an improvement to have a distinguished type of placeholder that could be incorporated into the template to control the choice.

The template pasting procedure thus far is as follows:

1. If there are placeholders in the template, choose one to be replaced.
2. If there are yet other placeholders in the template, choose one to be selected.
3. Replace the chosen placeholder, if any, with the target expression.
4. Substitute the resulting replacement expression for the target expression in the document.
5. Select the chosen placeholder, if any.

There remains the question of what to do in the last step if no placeholder was chosen to be selected. Remember, the critical issue is that *something* must be selected. One natural choice is as follows:

5′. Select the chosen placeholder. If none was chosen to be selected, select the replacement expression.

This allows the user to enter, for example, the expression

$$ax + b$$

entirely from the keyboard as follows:

$$\boxed{?} \xrightarrow{\mathbf{a}} \boxed{a} \xrightarrow{*} a\boxed{?} \xrightarrow{\mathbf{x}} a\boxed{x} \xrightarrow{space} \boxed{ax} \xrightarrow{+} ax+\boxed{?} \xrightarrow{\mathbf{b}} ax+\boxed{b}$$

The required sequence of keystrokes

a * x space **+ b**

bears at least some resemblance to the expression being entered. Only the * and the space seem superfluous.

The version of rule 5 that we actually use is the following:

5″. Select the chosen placeholder. If none was chosen to be selected, select the insertion point to the right of the replacement expression.

When combined with the next section's pasting procedure for insertion points, this yields extremely natural entry of linear notations for unary and binary operators. Nonlinear notations, such as exponents and fractions, are only slightly more troublesome to enter. On the other hand, operators that require more than two arguments unavoidably force a top-down style of entry. For example, although the user can wrap a definite-integral template around a previously entered integrand,

$$\boxed{x^2} \longrightarrow \int_?^? x^2 d\boxed{?}$$

he will then have to select and replace the remaining placeholders in turn.

6 Insertion Points

Every insertion point is either the left- or the right-hand side of an expression. For example, the expression $ax + b$ has five insertion points:

$$|a|\,x| + |b|$$

The leftmost insertion point is the left-hand side of three different expressions: a, ax, and $ax + b$. The next is the right side of a and the left side of x. The insertion points divide the displayed expression roughly as if it were a linear string of symbols.

It is not always the case that an insertion point neighboring an expression is identified with an insertion point neighboring the expression's first or last child. This can be because the two insertion points are at different horizontal positions on the screen, or because the expression's main operator arranges its arguments in a nonlinear way. Both cases are illustrated by the following:

$$|f|\,(|x|)| - \left|\frac{|1|}{|2|\,x|}\right|$$

The right side of $f(x)$ and the right side of its second argument, x, correspond to distinct insertion points because they are at different positions. The fraction does not at all resemble a string of symbols, so it is treated as if it were itself a single symbol. The overall expression thus forms a string of six symbols:

The numerator and denominator of the fraction form symbol strings of their own.

The user selects an insertion point by positioning the mouse and then clicking the left button. The system uses bounding boxes to determine the relevant symbol string:

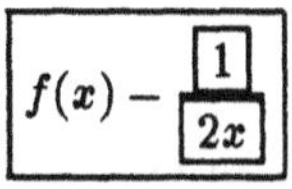

It then scans the sequence of insertion points for that string to find the one nearest the click position. The user can step the selection back and forth in the sequence with the left and right arrow keys.

All of this creates the illusion of linear expression structure for the user, but this structure is a fiction. The system must relate insertion points to the expression tree.

Internally, Kaava represents an insertion point as a combination of a path in the expression tree (a sequence of child indices starting from the root) and an indication of *left* or *right*. We have chosen to make the representation unique by preferring *right* over *left*, and a smaller expression over any expression containing it.

Externally, an insertion point identifies a node of the expression tree only ambiguously. This ambiguity makes insertion points quite awkward from a programming perspective. The same ambiguity is, in my opinion, the main justification for allowing the user to select insertion points.

Consider the insertion point to the right of the product xy:

$$x\ y|$$

If the user types a +, he will expect xy to become the left-hand argument:

$$x\ y| \xrightarrow{+} xy + \boxed{?}$$

If he types a ^ to indicate exponentiation, he will expect y to become the base:

$$x\ y| \xrightarrow{\hat{}} x y^{\boxed{?}}$$

Both cases involve wrapping a binary template around an expression, but the expressions is xy in the first case and y in the second. In each case, the choice happens to be the only one that directly produces a tree satisfying the structural schema, without the addition of parentheses.

So that the user can conveniently take control of the choice, hitting the space bar expands the selection from an insertion point to the smallest neighboring expression, preferably one on the left.

$$x\ y| \xrightarrow{space} x\boxed{y} \xrightarrow{+} x(y + \boxed{?})$$

One can hit it again to select the next larger expression.

$$x\ y| \xrightarrow{space} x\boxed{y} \xrightarrow{space} \boxed{xy} \xrightarrow{\hat{}} (xy)^{\boxed{?}}$$

Whether a template should be wrapped around anything at all when it is pasted at an insertion point is a matter of how the template appears to the user. Kaava leaves the decision up to the template designer. It replaces a placeholder with an expression to the left of the insertion point only if he has made the placeholder *left-sticky*. It replaces a placeholder with an expression to the right only if the placeholder is *right-sticky*. Stickyness is shown as a period on the left or right side of the question mark, as in the standard + template:

$$.?+?.$$

It is natural to make the base of an exponential left-sticky, but how to treat the exponent is a matter of taste. Both the base and the exponent are sticky in our standard exponential template:

$$.?^{?.}$$

In contrast, an integral template might include no sticky placeholders at all:

$$\int ?d?$$

Making the integrand right-sticky becomes an option, though, if we bind the template to a key that we expect the user to associate with the integral sign:

$$\$: \int ?.\, d?$$

When the template does not include an appropriately sticky placeholder, the system will have to actually insert it at the insertion point. Kaava does this with the aid of ordinary multiplication. It multiplies the original template by a sticky placeholder on either the left or the right to produce the effective template. In effect, it transforms the template x into one of the following:

$$.?x \quad x?.$$

Putting all of these ideas together into a pasting procedure is not easy. One of the compromises in Kaava is that the decision as to whether to wrap the template around an expression to the left or around an expression to the right (though not the choice of the particular expression) is made at the outset without considering the form of the template. If there is any neighboring expression at all on the left, Kaava chooses the left side. Here is the procedure for pasting at an insertion point:

1. If there is a placeholder that is sticky on the given side, mark it for replacement.
2. If there are yet other placeholders, choose one to be selected.
3. If no placeholder was marked for replacement, multiply the template on the given side by a placeholder marked for replacement.
4. Let the *target expression* be the smallest neighboring expression on the given side.
5. If the target expression is not the largest of the neighboring expressions on the given side, and if wrapping the template around it would not directly produce a document tree satisfying the structural schema, make the target expression's parent the new target. Repeat as many times as necessary.
6. Replace the placeholder marked for replacement with the target expression.
7. Substitute the resulting replacement expression for the target expression in the document.
8. If a placeholder was chosen to be selected, select it. Otherwise, depending on whether the given side is **L)** the left side or **R)** the right side, select the insertion point

   L) to the right of the replacement expression, or

   R) to the left of the placeholder that was replaced.

A few examples will wrap things up. The expression $ax+b$ of the previous section can now be entered as follows:

$$\boxed{?} \xrightarrow{\texttt{a}} a| \xrightarrow{\texttt{x}} a\,x| \xrightarrow{\texttt{+}} ax + \boxed{?} \xrightarrow{\texttt{b}} ax + b|$$

Let us say we want to start with a copy of this and produce the following quadratic:

$$ax^2 + bx + c$$

We select the insertion point to the right of the x and enter the exponent:

$$a\,x| + b \xrightarrow{\hat{}} ax^{\boxed{?}} + b \xrightarrow{2} ax^{2|} + b$$

Then we select the insertion point to the right of the b and enter the additional x and the third term:

$$ax^2 + b| \xrightarrow{\mathtt{x}} ax^2 + b\,x| \xrightarrow{+} ax^2 + bx + \boxed{?} \xrightarrow{\mathtt{c}} ax^2 + bx + c$$

We could also have entered the quadratic from scratch, in which case the sequence of keystrokes would have been as follows:

a x ^ 2 space space space **+ b x + c**

7 Conclusion

Usable formula editing interfaces are surprisingly complex. Only by distilling the essential ideas from our design attempts can we save these interfaces from creeping featurism.

This paper recommends the use of a literal expression representation corresponding closely to what the user sees on the screen. It also recommends the uniform, data-driven treatment of editing operations as template pasting. My recommendation regarding insertion points would depend on the available programming resources. The ambiguity that makes them useful in the first place also makes them rather difficult to implement.

The idea of sticky placeholders is one that I have not seen elsewhere. Although the semantics of pasting at an insertion point needs further refinement, Kaava's template notation will be a good foundation for this work.

Finally, I would like to reiterate that structure editing as described in this paper has a competitor: incremental parsing. I hope that the paper is a step towards an eventual understanding of the tradeoffs between these two approaches.

Acknowledgments

This work was supported by the Academy of Finland. Kim Nyberg, Harri Pasanen, Kenneth Oksanen, Tero Mononen, and Harri Hakula assisted in developing Kaava as it now stands. I especially want to thank Harri Pasanen, whose thesis has significantly influenced the exposition in this paper.

References

1. Dennis Arnon, Richard Beach, Kevin McIsaac, and Carl Waldspurger. CaminoReal: An interactive mathematical notebook. Technical Report EDL-89-1, Xerox PARC, 1989.
2. Ron Avitzur. Suggestions for a friendlier user interface. In *Proc. DISCO '90: Symposium on the Design and Implementation of Symbolic Computation Systems*, pages 282–283, 1990.
3. Frame Technology Corporation, San Jose, California. *Using FrameMath*, 1989.

4. Gail E. Kaiser and Elaine Kant. Incremental parsing without a parser. *Journal of Systems and Software*, 5:121–144, 1985.
5. Paracomp, San Francisco, California. *Milo User's Guide*, 1988.
6. Harri Pasanen. Highly interactive computer algebra. Master's thesis, Helsinki University of Technology, Department of Computer Science, 1992. Available as Technical Report TKO-C52.
7. Carolyn Smith and Neil Soiffer. MathScribe: A user interface for computer algebra systems. In *Proc. 1986 Symposium on Symbolic and Algebraic Computation*, pages 7–12. Association for Computing Machinery, 1986.
8. Neil Soiffer. *The Design of a User Interface for Computer Algebra Systems*. PhD thesis, University of California at Berkeley, 1991.

Algebraic Simplification of Multiple-Valued Functions

Russell Bradford

School of Mathematical Sciences
University of Bath
Claverton Down
Bath BA2 7AY

1 Abstract

Many current algebra systems have a lax attitude to the simplification of expressions involving functions like log and $\sqrt{}$, leading to the ability to "prove" equalities like $-1 = 1$ in such systems. In fact, only a little elementary arithmetic is needed to devise what the correct simplifications should be. We detail some of these simplification rules, and outline a method for their incorporation into an algebra system.

2 Introduction

Algebra systems, like pocket calculators, ought to be correct. Many, *unlike* most pocket calculators, are not. Typically they will simplify $\sqrt{a/b}$ to $\sqrt{a}/\sqrt{b}$ on one line, but $\sqrt{1/-1}$ to $\sqrt{-1}$ on the next, not noticing the contradictory results they have just given. Some have tried to avoid the question by refusing to simplify *any* square roots, thus missing such cases as $\sqrt{i/4}$ which is indeed $\sqrt{i}/\sqrt{4}$. This simplification is valid when $\arg a \geq \arg b$ (the complex argument), which covers the "usual" case of both a and b real and positive.

The much-touted suggestion that $\sqrt{z^2} = |z|$, the absolute value of z, only serves to point up the muddy thinking on this point: it is certainly not true that $\sqrt{i^2} = |i|$—there is a bias toward the real line that appears throughout the subject of algebraic simplification. Many simplification "rules," such as $\ln(ab) = \ln a + \ln b$, are given that are certainly true when a and b are both real and positive, but are false in many wider senses, such as a and b both negative. In fact, some of the well known rules are false in *every* wider sense! For example, $\ln(1/b) = -\ln b$ when b is positive real, but $\ln(1/b) = 2\pi i - \ln b$ for every other complex b.

In the same vein, it is often said that $\arccos \cos z = z$, even though it is clear that this equality is only true for a limited set of values of z. When on their best behaviour, algebra systems (and people) will recognise the last equality as false, but few will have qualms over statements such as $(1/b)^c = 1/b^c$.

See Stoutemyer [2] for many more examples, and a discussion of what the user expects as an answer to such simplifications in contrast to what algebra systems currently offer (and, further, whether either is the correct answer!). Kahan [1] also has much to say on the subject.

3 Principles of Principals

Throughout this paper we shall take the complex argument function to run from 0 (inclusive) to 2π (exclusive). Also, when considering any multiple-valued function we shall be taking its principal value (the value with least argument). This choice of the argument and principal value makes the following exposition simpler—of course, any other self-consistent choice could be made with equal validity. Thus, we have $\sqrt{-2i} = (-2i)^{1/2} = -1 + i$, and $(-8)^{1/3} = 1 + \sqrt{3}i$.

It is a commonly felt belief that evaluation should commute with simplification. Thus substituting $x = 2$ in the simplification $x^2 + 2xy + y$ of $(x+y)^2$ should result in the same value as simplifying the result of substituting $x = 2$ in $(x+y)^2$. This seems clear enough, but consider substituting $z = 2\pi i$ into $\ln e^z$ (to get 0), in contrast with simplifying $\ln e^z$ to z, and then substituting (to get $2\pi i$).

Of course, the "simplification" was incorrect, the problem being that ln is a many-valued function, and the choice of branch is of utmost importance if we are to have any hope of being correct. It is this choice of branch that causes many algebra systems to fall foul of mathematics: either in the choice of the wrong branch, or a refusal to choose any branch at all.

We define $\operatorname{adj} z$ for the complex value z by $\ln(e^z) = z + \operatorname{adj} z$. In essence, $\operatorname{adj} z$ is the discrepancy between the "expected" value of ln, and the principal value actually obtained. It is a matter of simple calculation to find that $\operatorname{adj} z = -2\pi i \lfloor \operatorname{Im} z/2\pi \rfloor$ (where $\lfloor x \rfloor$ is the largest integer not greater than x). Thus the adjustment increases as a step function of size 2π with the imaginary part of z.

$$\ln(e^z) = z - 2\pi i \left\lfloor \frac{\operatorname{Im} z}{2\pi} \right\rfloor .$$

Hence, we correctly see that $\ln e^{2\pi i} = 2\pi i - 2\pi i \times 1 = 0$, and not $2\pi i$ as a naïve cancellation $\ln e^z = z$ might "prove."

4 Simple Properties

Here z and w are arbitrary complex, x is real, and n is integral.

$$\operatorname{adj} \operatorname{adj} z = -\operatorname{adj} z \qquad \operatorname{adj}(\operatorname{adj} z + \operatorname{adj} w) = -\operatorname{adj} z - \operatorname{adj} w$$
$$\operatorname{adj}(z + \operatorname{adj} w) = \operatorname{adj} z - \operatorname{adj} w \qquad \operatorname{adj}(z + 2n\pi i) = \operatorname{adj} z - 2n\pi i$$
$$\operatorname{adj} z = \operatorname{adj}(i \operatorname{Im} z) \qquad \operatorname{adj}(z + x) = \operatorname{adj} z$$
$$e^{n \operatorname{adj} z} = 1 \qquad \operatorname{adj} \ln z = \operatorname{adj}(i \arg z) = 0, \text{ as } \arg z < 2\pi$$
$$\operatorname{adj} \operatorname{Re} z = \operatorname{adj} \operatorname{Im} z = \operatorname{adj} x = 0$$

5 Log Simplifications

As a simple first example, consider when $\ln(ab) = \ln a + \ln b$. Now, $\ln(-1.-1) = \ln(1) = 0$, but $\ln(-1) + \ln(-1) = i\pi + i\pi = 2\pi i$, so this rule does not hold over the entire complex plane.

We have

$$\begin{aligned}\ln(ab) &= \ln(e^{\ln a}e^{\ln b})\\ &= \ln(e^{\ln a+\ln b})\\ &= \ln a + \ln b + \mathrm{adj}(\ln a + \ln b)\end{aligned}$$

Here we have used the fact (proven by analysis) that $e^z e^w = e^{z+w}$ for all complex z and w.

Taking this we get $\mathrm{adj}(\ln a + \ln b) = -2\pi i\lfloor \mathrm{Im}(\ln a + \ln b)/2\pi\rfloor = -2\pi i\lfloor(\arg a + \arg b)/2\pi\rfloor$. Now, $0 \le \arg z < 2\pi$, so we see that $\lfloor(\arg a + \arg b)/2\pi\rfloor = 0$ if either $\arg a = 0$, or $\arg b = 0$, i.e., either of a and b are real and positive. Thus, our usage of log tables in school to do multiplications of positive numbers was correct! In detail,

$$\left\lfloor \frac{\arg a + \arg b}{2\pi} \right\rfloor = \begin{cases} 0 & \text{when } \arg a + \arg b < 2\pi \\ 1 & \text{when } \arg a + \arg b \ge 2\pi \end{cases}$$

thus

$$\ln(ab) = \begin{cases} \ln a + \ln b & \text{when } \arg a + \arg b < 2\pi \\ \ln a + \ln b - 2\pi i & \text{when } \arg a + \arg b \ge 2\pi \end{cases}$$

Similarly, for the quotient:

$$\begin{aligned}\ln(a/b) &= \ln a - \ln b + \mathrm{adj}(\ln a - \ln b)\\ &= \begin{cases} \ln a - \ln b & \text{when } \arg a \ge \arg b \\ \ln a - \ln b + 2\pi i & \text{when } \arg a < \arg b \end{cases}\end{aligned}$$

So $\ln(a/b) = \ln a - \ln b$ when b is real and positive; or if both a and b are real, and have the same sign; or if a is a positive real multiple of b.

Special cases of the above are:

$$\ln(-b) = \begin{cases} \ln b + i\pi & \text{when } 0 \le \arg b < \pi \\ \ln b - i\pi & \text{when } \pi \le \arg b < 2\pi \end{cases}$$

so we see that $\ln(-b) = \ln b + i\pi$ when $b > 0$, but $\ln(-b) = \ln b - i\pi$ when $b < 0$.

Again,

$$\ln(1/b) = \begin{cases} -\ln b & \text{when } \arg b = 0 \\ 2\pi i - \ln b & \text{when } \arg b > 0 \end{cases}$$

In this case, the "obvious" simplification is correct only when b is real and positive!

Another log expression:

$$\ln(a^b) = b\ln a + \mathrm{adj}(b\ln a).$$

So $\ln(a^b) = b\ln a$ if $a > 0$ and b is real; or if b is real and $0 \le b \le 1$. The latter is true since for these values of b, $0 \le b\,\mathrm{Im}\ln a < 2\pi$, whence $\mathrm{adj}(b\ln a) = 0$.

6 Exp Simplifications

Moving on to exponentiation:

$$(a^b)^c = a^{bc} e^{c\,\mathrm{adj}(b \ln a)}.$$

Hence, $(a^b)^c = a^{bc}$ if $a > 0$, and b is real; or if c is integral. The latter is true since $c\,\mathrm{adj}(b \ln a)$ is an integral multiple of $2\pi i$ whenever c is an integer, whence $e^{c\,\mathrm{adj}(b \ln a)} = 1$. So raising to an integral power is always "safe," but in general, there is an extra factor.

A useful special case is

$$(a^2)^{1/2} = \begin{cases} a & \text{when } 0 \le \arg a < \pi \\ -a & \text{when } \pi < \arg a < 2\pi \end{cases}$$

Square roots simplify in the "obvious" manner only in the upper half of the complex plane (with this choice of arg). Thus, $\sqrt{(1+i)^2} = \sqrt{2i} = 1 + i$, but $\sqrt{(1-i)^2} = \sqrt{-2i} = -1 + i$. This special case is the source of the $\sqrt{z^2} = |z|$ confusion: because this simplification is true for all real numbers, it is commonly assumed to be true for all complex numbers.

Similarly

$$(ab)^c = a^c b^c e^{c\,\mathrm{adj}(\ln a + \ln b)}.$$

So now, $(ab)^c = a^c b^c$ if either of a and b are real positive; or if c is integral.

In the case of square roots:

$$\sqrt{ab} = (ab)^{1/2} = \begin{cases} \sqrt{a}\sqrt{b} & \text{when } \arg a + \arg b < 2\pi \\ -\sqrt{a}\sqrt{b} & \text{when } \arg a + \arg b \ge 2\pi \end{cases}$$

And for division

$$\begin{aligned}(a/b)^c &= \frac{a^c}{b^c} e^{c\,\mathrm{adj}(\ln a - \ln b)} \\ &= \begin{cases} (a^c/b^c) & \text{when } \arg a \ge \arg b \\ (a^c/b^c) e^{2c\pi i} & \text{when } \arg a < \arg b \end{cases}\end{aligned}$$

So $(a/b)^c = a^c/b^c$ if b is positive real; or if c is integral.

A special case:

$$\begin{aligned}(1/b)^c &= (1/b)^c e^{c\,\mathrm{adj}(-\ln b)} \\ &= \begin{cases} 1/b^c & \text{when } \arg b = 0 \\ e^{2c\pi i}/b^c & \text{when } \arg b > 0 \end{cases}\end{aligned}$$

Again, the "obvious" simplification is correct in only a limited set of examples, viz., when b is real and positive; or if c is integral.

7 Trig Simplifications

Trigonometric functions tend to be treated more carefully in algebra systems, as they are periodic along the real line, and so their inverses are quite clearly multiple valued. This is in contrast to the exponential function, whose periodicity is often overlooked.

Simplification rules for the trigonometric functions can be derived from the above as they can be expressed in terms of exps and logs, thus $\cos z = (e^{iz} + e^{-iz})/2$, and $\arccos w = -i \ln(w + i\sqrt{1-w^2})$, so

$$\begin{aligned} \arccos\cos z &= -i\ln(\cos z + i\sqrt{\sin^2 z}) \\ &= -i\ln(\cos z + i\sin z) \qquad \text{when } 0 \le \arg\sin z < \pi \\ &= -i\ln e^{iz} \\ &= -i(iz - \operatorname{adj}(iz)) \\ &= z - 2\pi\lfloor \operatorname{Re} z/2\pi \rfloor . \end{aligned}$$

In the case that $\pi \le \arg\sin z < 2\pi$, (e.g., when $\sin z < 0$), so $\sqrt{\sin^2 z} = -\sin z$, we find

$$\begin{aligned} \arccos\cos z &= -i\ln(\cos z - i\sin z) \\ &= -i\ln e^{-iz} \\ &= -z - 2\pi\lfloor \operatorname{Re}(-z)/2\pi \rfloor . \end{aligned}$$

Thus we have

$$\arccos\cos z = \begin{cases} z - 2\pi\lfloor \operatorname{Re} z/2\pi \rfloor & \text{when } 0 \le \arg\sin z < \pi, \\ -z + 2\pi\lceil \operatorname{Re} z/2\pi \rceil & \text{when } \pi \le \arg\sin z < 2\pi \end{cases}$$

In the case when x is real, $0 \le \arg\sin x < \pi$ when $0 < \sin x \le 1$, which holds when $2n\pi < x < (2n+1)\pi$, for integral n. So

$$\arccos\cos x = \begin{cases} x - 2\pi\lfloor x/2\pi \rfloor & \text{when } 2n\pi \le x < (2n+1)\pi, \\ -x + 2\pi\lceil x/2\pi \rceil & \text{when } (2n+1)\pi \le x < 2n\pi \end{cases}$$

This is a sawtooth function:

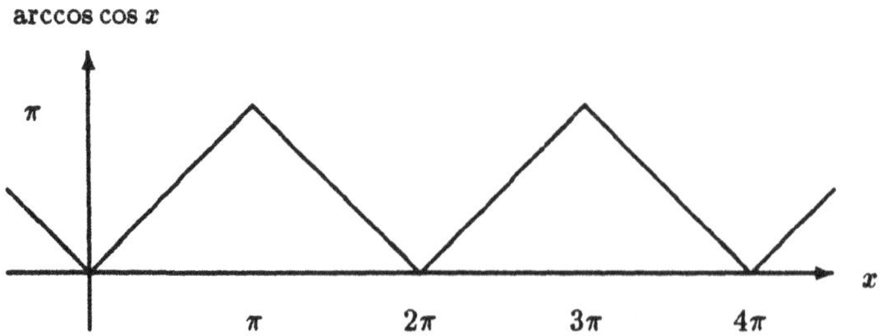

Similar simplifications hold true for the rest of the trigonometric and hyperbolic functions, e.g.,

$$\operatorname{arccosh}\cosh z = \begin{cases} z - 2\pi i\lfloor \operatorname{Im} z/2\pi \rfloor & \text{when } 0 \le \arg\sinh z < \pi, \\ -z + 2\pi i\lceil \operatorname{Im} z/2\pi \rceil & \text{when } \pi \le \arg\sinh z < 2\pi \end{cases}$$

In the case of the hyperbolics, the restrictions to the real line are particularly simple, namely $\operatorname{arcsinh}\sinh x = x$, $\operatorname{arctanh}\tanh x = x$, and $\operatorname{arccosh}\cosh x = x$, if $x \ge 0$, but $\operatorname{arccosh}\cosh x = -x$, if $x < 0$.

8 Application

Next we consider the application of the above rules to practical situations.

The general user of an algebra system is would prefer not to see a complicated answer consisting of several subcases dependent on the size of the argument of some difficult expression (say), but would rather see the simple answer that (sometimes) arises in the special case of interest, e.g., when x is real, and y is real and non-negative.

Thus we have the need to be able to attach attributes to symbols that describe properties of those symbols. Thus we might wish to indicate that x is real and y is non-negative by properties such as `Re` and `>0`. However, such symbols soon become unwieldy, and we need a definite algebra for them, such as described in [3] and [4]. In these papers Weibel and Gonnet describe a general framework for a lattice of properties of objects, which we can use here.

9 Lattice

As we have seen above, knowing the complex argument of an expression is essential for correct choice of branch. Keeping a full record of the possible ranges of the argument for expressions is theoretically impossible, and practically very hard even for a restricted set of functions. Thus we suggest a compromise that is sufficiently detailed to cover most of the "usual cases" (e.g., x is real, w is negative, z has positive imaginary part), but is not so complicated as to be intractible.

For each of the real and imaginary parts of an expression we classify the value as being positive (> 0), negative (< 0), or zero ($= 0$), or some combination thereof (≥ 0, ≤ 0, $\neq 0$). This gives us a lattice based on a cross product of two of the following:

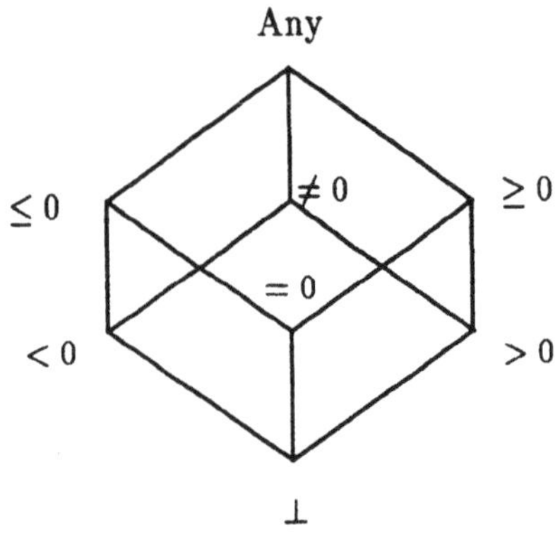

At the bottom of the combined lattice we have $\perp$, the indicator of an impossible combination of attributes, and at the top we have the entire complex plane **C**, indicating we know nothing about the number whatsoever. A typical attribute of a symbol z will be, say, $(\geq 0, \geq 0)$ to indicate the range $0 \leq \arg z \leq \pi/2$, or $(> 0, = 0)$ to denote z real and positive.

Next, we can use a simple calculus of con- and disjunctions to combine these, say if it is known that z has both $(\leq 0, any)$ and $(= 0, > 0)$, then we may deduce that z has $(= 0, > 0)$. Similarly, if z has either $(\leq 0, = 0)$ or $(\neq 0, > 0)$, then z has $(any, \geq 0)$.

We have implemented the above by using simple bit-vectors: for each of the real and imaginary parts a triple of bits is used to represent the sign. The first bit is set if the value is possibly negative, the second is set if the value is possibly zero, the third if the value is possibly positive. Thus, say, < 0 is the vector `100`, whereas `101` is $\neq 0$. The combination `000` is $\perp$, and `111` represents the undetermined value. Logical conjunctions and disjunctions of these translate directly to logical ands and ors of the bit patterns, together with the note that if either the real or imaginary parts are $\perp$, then both are.

10 Calculus

Having properties on symbols alone is not enough—we need to be able to assign properties to expressions, for example can we deduce anything about the sign or argument of $\sin z$ given the sign or argument of z?

Clearly, in this case, starting from "z is real" (viz., $(any, = 0)$) we can get to "$\sin z$ is real." More is true, but our classification scheme is too coarse to specify this in more detail.

In [3, 4] there is a complete calculus of how to compute or bound the position on the lattice of $F(z)$ given information (possibly incomplete) on both F and z. Thus, for sin, we can have a map $(any, = 0) \rightarrow (any, = 0)$, indicating that it maps reals to reals; or, for $+$, we might have a map $(> 0, = 0) \times (\geq 0, = 0) \rightarrow (\geq 0, = 0)$, indicating that the sum of a positive number and a non-negative number is positive.

The position computed may be (any, any) (i.e., the whole complex plane) in the worst case, but can give much useful information, particularly about the reality or otherwise of a number, since many elementary functions map reals to reals.

11 Synthesis

Putting the above together, we have simplification rules for certain expressions, and a mechanism for determining when to apply these rules.

An example. Given x is real and non-negative, deduce that $\sqrt{(x+1)^2} = x+1$. The information we have is that x is $(\geq 0, = 0)$, and the signature of $+$. Now 1 is $(> 0, = 0)$, and the sum of a $(\geq 0, = 0)$ and a $(> 0, = 0)$ is a $(> 0, = 0)$. Thus $x+1$ is real and positive. This implies that $\arg(x+1)$ is 0, so that we may apply the simplification rule $\sqrt{a^2} \rightarrow a$. Hence, $\sqrt{(x+1)^2} = x+1$.

Now, if we tried to simplify $\sqrt{(x-1)^2}$ we could only deduce that $x-1$ is real, and we could not apply the same cancellation. Indeed, for $x = 5$, $x-1 = 4$, and $\sqrt{(x-1)^2} = 4$, but for $x = 1/2$, $x-1 = -1/2$, whereas $\sqrt{(x-1)^2} = 1/2$.

Given x is real, and z is arbitrary, what can we deduce about $\sqrt{z\cosh^2 x}$? Well, x has $(any, = 0)$ implies $\cosh x$ has $(> 0, = 0)$, whence $\cosh^2 x$ has $(> 0, = 0)$. Thus, $\cosh^2 x$ is real and positive, so we may apply $\sqrt{ab} \rightarrow \sqrt{a}\sqrt{b}$, to deduce $\sqrt{z\cosh^2 x} = \sqrt{z}\sqrt{\cosh^2 x}$. Next, we can apply $\sqrt{a^2} \rightarrow a$ for real positive a ($\cosh x$ is real and postive) to get $\sqrt{z\cosh^2 x} = \sqrt{z}\cosh x$. This example is in contrast to $\sqrt{z\cos^2 x}$ where we can only get as far as $\sqrt{z\cos^2 x} = \sqrt{z}\sqrt{\cos^2 x}$, as $\cos x$ may be negative.

12 Conclusion

We have presented a way to avoid some of the worst errors made in the simplification of expressions containing multiple-valued functions. By a type inference on the value of the expression we can determine when a certain simplification is valid, or determine that we cannot simplify an expression without the possibility of introducing an error.

It may be that we cannot deduce much information about the range of values of a particular expression, but this is in itself useful, as it indicates that we may not blindly apply a naïve simplification rule, but must present the full set of possible values for the expression.

For the future, we would like to increase the resolution of the lattice, say split the line into $(-\infty,-1)\cup\{-1\}\cup(-1,0)\cup\{0\}\cup(0,1)\cup\{1\}\cup(1,\infty)$, as this will allow us to pinpoint functions that use the range 0 to 1, a common interval. For example, log is real and negative on $(0,1)$, zero on $\{1\}$, real and positive on $(1,\infty)$. And cos maps reals to $[0,1]$, and so on.

However, this increases the work needed to describe the elementary functions immensely: just consider what is necessary to precompute all the possible combinations for $+$. However, some of this can be avoided by compute on need, or deduction by means of the calculus.

References

1. W Kahan. The Interface between Symbolic and Numberic Computation. seminar notes, IBM Oberlech, Austria, July 1991.
2. David R Stoutemyer. Crimes and Misdemeanors in the Computer Algebra Trade. *Notices of the AMS*, 38(7):778–785, September 1991.
3. Trudy Weibel and G H Gonnet. An Algebra of Properties. Technical Report 157, ETH Zürich, Institut für Wissenschaftliches Rechnen, ETH-Zentrum, CH-8092, Zurich, Switzerland, April 1991.
4. Trudy Weibel and G H Gonnet. An Algebra of Properties. In Stephen M Watt, editor, *Proceedings of the International Symposium on Symbolic and Algebraic Computation (ISSAC 91)*, pages 352–359. ACM Press, July 1991.

In-place Arithmetic for Polynomials over $\mathbf{Z}_n$

Michael Monagan

Institut für Wissenschaftliches Rechnen
ETH-Zentrum, CH-8092 Zürich, Switzerland
monagan@inf.ethz.ch

Abstract. We present space and time efficient algorithms for univariate polynomial arithmetic operations over **Z** mod n where the modulus n does not necessarily fit into is not a machine word. These algorithms provide the key tools for the efficient implementation of polynomial resultant gcd and factorization computation over **Z**, without having to write large amounts of code in a systems implementation language.

1 Background

This paper reports a solution to a dilema we faced during the design and implementation of certain crucial operations in the Maple system [15] namely, computing polynomial resultants, greatest common divisors and factorization in $\mathbf{Z}[x]$.

The efficient implementation of polynomial greatest common divisors (Gcds) is perhaps the most single important part of a general purpose computer algebra system. Gcd computation is the bottleneck of many operations. This is because any calculations which involve rational operations will require Gcd computations in order to reduce fractions to lowest terms. For example, in solving a system of equations with polynomial coefficients, polynomial Gcd calculations will be needed to simplify the solutions. Whereas we can use Euclids algorithm to compute integer Gcds relatively efficiently, compared with integer multiplication and division, the efficient computation of polynomial Gcds is much more difficult. And although the classical algorithms for multiplying and dividing polynomials are fine for most practical calculations, the use of Euclidean based Gcd algorithms results in a phenomenon known as "intermediate expression swell" which causes many intermediate calculations to "blow up". Research on Gcd computations in the 1970's and 1980's [9, 19, 21, 7, 12] led to efficient algorithms for the computation of polynomial Gcds. However, the efficient implementation of these algorithms is quite difficult. The modular based algorithms require efficient computation over the integers mod n. Efficiency is lost in a systems implementation language like C and Lisp, and even more so in an interpreted language.

The efficient implementation of polynomial resultants is not as important as polynomial Gcds. And it has become less important since one of the main applications of resultants, namely in solving systems of polynomial equations, has largely been superceded with the Grobner basis approach [5, 10]. However, it has been our experience that some hard problems can only be solved by clever application of resultants [16]. Thus it is still useful to have an efficient implementation of polynomial resultants. The most efficient algorithms for resultants for dense polynomials are those based on modular methods [9, 4].

Polynomial factorization is another important facility in a computer algebra system. Various algorithms require polynomial factorization, such as computing with algebraic numbers. But also, factorization is one way the user can try to "simplify" an expression. Factored polynomials are usually smaller in size than expanded polynomials.

In 1986, in Maple version 4.2, these operations were implemented as follows. Resultants were computed using the sub-resultant algorithm [9, 4]. Gcds were computed using the Heuristic algorithm GCDHEU [7]. And polynomial factorization over $\mathbf{Z}$ was done using the Berlekamp-Hensel procedure [13, 10].

For dense polynomials, it had been known for some time that the modular methods of Collins [9] were superior to the sub-resultant algorithm for polynomial Resultant and Gcd computation. However, an efficient implementation requires an efficient implementation of polynomial arithmetic in $\mathbf{Z}_p[x]$ where p will be a machine word size prime, and an efficient implementation of the Chinese remainder algorithm for combining images. Maple did not use this approach because firstly, the implementation in Maple would be completely interpreted, and secondly, Maples implementation of the sub-resultant algorithm for computing Resultants and the GCDHEU algorithm for Gcds was quite competitive for many problems. Essentially, Maple was able to "piggy back" off its efficient implementation of multi-precision integer arithmetic and univariate polynomial arithmetic over $\mathbf{Z}$. But the implementation of the Berlekamp-Hensel procedure for factorization requires polynomial arithmetic and linear algebra over $\mathbf{Z}_p$ and polynomial arithmetic over $\mathbf{Z}_{p^k}$. This was all implemented in Maple. Maple was very slow at factoring over finite fields and consequently also slow at factoring over $\mathbf{Z}$. For some time this was not seem as a serious problem because the competition, namely REDUCE and MACSYMA, also had and still have relatively slow univariate polynomial factorization packages. However, just how slow Maple was, and REDUCE and MACSYMA for that matter, became apparent when SMP timings were reported. To give one comparison, the landmark SIGSAM problem No. 7 [11], which includes a factorizaton of a polynomial of degree 40 with over 40 digit coefficients could be done in about a minute on a Vax 11/780 using SMP but took Maple well over an hour.

So, how do we improve the performance of factorization in $\mathbf{Z}[x]$ in Maple? We recognize that Maple is too slow because it is interpreted. Thus what we have to do is to implement the polynomial factorization package and the tools it requires in C. There will be some economy of code since many of the same tools needed for factorization can also used to implement Gcds and resultants efficiently. However, anyone who has implemented a polynomial factorization package knows that this is a formidable task. The amount of coding is considerable. For instance, it is said that when Arthur Normal first implemented polynomial factorization in REDUCE, the overall size of the whole system doubled! I have seen the the SMP implementation. I do not know how many lines of code it was but it was a stack of paper about 1 inch think. This solution presents us with a major dilema in Maple. We have for years tried to avoid coding in C and instead to code in the Maple programming language. The advantage is clear. Maple code is easier to write, read, debug and maintain. We also wanted to keep the Maple kernel, that is, the part of the Maple system that is written in C, as small as possible. This was done primarily so that there would be more storage available for data. However, there are other clear advantages in maintenance and portability. But, Maple is interpreted and hence some operations will execute slowly. Operations with small integers and floating point numbers excute particularly slowly in Maple compared with compiled C. Hence also polynomial arithmetic and linear algebra over $\mathbf{Z}_n$ executes slowly. One is tempted to simply code all these operations in C. Thus the dilema we faced was, how do we get an efficient polynomial factorization package without writing tens of thousands of lines of C code.

To summarize our finding here, the answer appears to be that one can have an efficient implementation of polynomial resultants, Gcds and factorizaton over $\mathbf{Z}$ if the system supports efficient arithmetic in $\mathbf{Z}_n[x]$. It is not necessary to code everything in C. Specifically, only addition, subtraction, multiplication, division, quotient, remainder, evaluation, interpolation Gcd and resultant in $\mathbf{Z}_n[x]$ need to be coded in C. Factorization over $\mathbf{Z}_p$, and resultants, Gcds and factorization over $\mathbf{Z}$ can then be coded efficiently in terms of these primitives.

Our other major finding is that if we focus on coding these primitives for $\mathbf{Z}_n[x]$ carefully, in particular, multiplication quotient and remainder, we can get modest improvements of factors of 3 to 5 by being careful about the way we handle storage and reduce modulo $\mathbf{Z}_n$. The result is a package which is partially written in C and mostly written in Maple. That amount of C code that we wrote is 1200 lines. This investment gives us fast implementations of the three key operations mentioned previously, and also factorization over $\mathbf{Z}_p$.

Since then we have used these primitives to improve the performance of other operations. We have implemented the multiple modular method of Collins [9] for bivariate polynomial resultants and Gcds. Also we have used the fast arithmetic for $\mathbf{Z}_p[x]$ to represent large finite fields $\mathrm{GF}(p^k) = \mathbf{Z}_p[x]/(a)$ where $a \in \mathbf{Z}_p[x]$ is irreducible. One then obtains a fast implementation of univariate polynomial arithmetic, including Gcds, resultants and factorization over $\mathrm{GF}(p^k)$.

We note that the computationally intensive steps of the two fast Gcd algorithms compared by Smedley [18] for computing Gcds of univariate polynomials over an algebraic number field which is a simple algebraic extension of $\mathbf{Q}$ also use our codes. The first method [17] is a heuristic method that reduces the problem to a single Gcd computation over $\mathbf{Z}_n$ for a large possibly composite modulus n. The second method [14] is a multiple modular method. Gcds are computed over $\mathbf{Z}_{p_i}[x]/(a)$ for word size prime moduli p_i and combined by application of the Chinese remainder theorem.

2 Introduction

How does one implement efficient univariate polynomial arithmetic over $\mathbf{Z}$, in particular the operations Gcd, resultant, and factorization? The fastest known practical method for computing Gcds and resultants is the dense modular method. For a full description of this method we refer the reader to [9, 10]. Briefly, given $a, b \in \mathbf{Z}[x]$, one computes the $\mathrm{Gcd}(a, b)$ (the $\mathrm{resultant}(a, b)$) modulo primes $p_1, \ldots, p_n$ and then combines the images using the Chinese remainder theorem. There are many details but this is the basic idea. In order to do this most efficiently, one chooses the primes $p_1, \ldots, p_n$ to be the biggest primes that fit into a machine word, so that one can use machine arithmetic directly for calculations in $\mathbf{Z}_{p_i}$. One also needs an efficient implementation of Chinese remaindering for combining the image Gcds or resultants.

This can be implemented efficiently by representing the polynomials over $\mathbf{Z}_{p_i}$ as arrays of machine integers and using the Euclidean algorithm for computing the Gcd and resultant. However, an implementation in C will lose efficiency because in order to multiply in $\mathbf{Z}_p$ with remainder, one can only use half a machine word as otherwise the product will overflow and the leading bits will be lost. Even though almost all hardware has instructions for multiplying two full word numbers and getting the two word result, these instructions are not accessible from C. Lisp implementations lose efficiency because their representation of integers is special. Some bits may be used for special purposes and there is an overhead for arithmetic operations. Note, in AXIOM there is the additional overhead of function calls. Basically, for various reasons, machine efficiency is lost.

Note: if one wants to handle multivariate polynomials, one will also need efficient implementations in C of polynomial evaluation and interpolation. The implementation of the Gcd and resultant computation over $\mathbf{Z}$ can be implemented in the high level language, and even if interpreted, the overhead be relatively insignificant.

Polynomial factorization is considerably more difficult. Given an efficient implementation of polynomial addition, multiplication, quotient and remainder, Gcd, over $\mathbf{Z}_p$ one can write an efficient procedure for factorization of univariate polynomials over $\mathbf{Z}_p$ using the Cantor-Zassenhaus distinct degree factorization algorithm [6]. The bottleneck of this computation is computing the remainder of a^n divided b for large n where $a, b \in \mathbf{Z}_p[x]$. This can be done efficiently using binary powering with remainder and requires only multiplication and remainder operations in $\mathbf{Z}_p[x]$. Note, the efficiency of this procedure can be improved slightly if one can square a polynomial efficienly. This is worth doing. We found that it saves about 15% overall. The next part of the Berlekamp-Hensel procedure for factorization is to lift the image factors using P-adic lifting (Hensel lifting) from $\mathbf{Z}_p$ to $\mathbf{Z}_{p^k}$ until p^k bounds twice the largest coefficient that could appear in any factor over $\mathbf{Z}$. The details of Hensel lifting can be found in [10, 13]. From our view in this paper, what this means is that one must be able to do polynomial arithmetic over $\mathbf{Z}_{p^k}$ efficiently. In particular, multiplication, quotient and remainder. During the lifting, the modulus p^i will eventually exceed the word size and one is forced to use multi-precision arithmetic. How can we efficiently multiply and divide over $\mathbf{Z}_{p^k}$? The next section of this paper addresses this problem. The factorization problem is then completed by trying combinations of the lifted factors to see if they divide the original input. Again, the details are many. Good references include the texts [10, 13].

3 In Place Algorithms

In this section we design an efficient environment for computing with univariate polynomials over the finite rings $\mathbf{Z}_n$ where n is too large to fit in a machine word, and $\mathbf{Z}_p[x]/(a)$ where $a \in \mathbf{Z}_p[x]$ and p here is word size prime. Let R denote either of these finite rings. Our implementation uses classical algorithms for arithmetic in R, since, in almost all cases, the size of the rings will not be large enough to warrant the use of asymptotically fast algorithms. Our implementation also uses classical algorithms for $R[x]$, since again, for most cases, the degree of the polynomials will not be large enough for asymptotically fast algorithms to win out. We are going to optimize the implementation at the level of storage management and data representation.

In a *generic* implementation of univariate polynomial arithmetic over R (as one would find in AXIOM for example) each arithmetic operation in R implies the creation of a new objects. Each new object created means storage management overhead. For example, to multiply a, b in $\mathbf{Z}_n$ we would first compute $c = a \times b$ then the result $c \bmod n$. In doing so several pieces of storage will be allocated which will later have to be garbage collected. We describe a more efficient strategy for computing in $R[x]$ which eliminates this overhead. The idea is to exploit the fact that unlike arithmetic over an arbitrary ring e.g. $\mathbf{Q}$, the storage required for arithmetic operations over R is bounded a priori since R is a finite ring.

We exploit this by coding arithmetic to run *in-place*. For arithmetic operations in $R[x]$ we will either pre-allocate the storage needed for the entire operation or, write out the answer as we go using an *on-line* algorithm. The algorithms given for $R[x]$ all allocate linear total storage in the size of the inputs, assuming the inputs are dense, which they usually are. In the case of polynomial multiplication and division, we can do this in the space required to write down the answer plus a constant number of scratch registers for arithmetic in R. Thus our algorithms are space optimal up to lower order terms. Another significant improvement can be obtained by allowing values to accumulate before reducing modulo n (or a) hence eliminating expensive operations.

The assumption here is that storage management; that is the overhead of allocating storage for each operation, and garbage collection is significant compared with the arithmetic operations involved. The overhead of storage management is surely the main reason why numerical software systems are inherently faster than symbolic algebra systems. This is simply because the primitive objects being manipulated in numerical software systems, namely floating point numbers, have fixed size. Because of this, immediate storage structures can be used for vectors and matrices of floating point numbers. Storage management is trivial in comparison. Now the size of objects in $\mathbf{Z}_n$ and $\mathbf{Z}_p[x]/(a)$ is not fixed. It is parameterized by n, p and $deg(a)$. However, unlike $\mathbf{Q}$, the size depends only on the domain, not on the values of the domain. The difference is significant. For operations over Q the size of the result will depend on additional parameters such as the degree of a polynomial. Although it may be possible to bound the storage needed for arithmetic over $\mathbf{Q}$ and design in-place algorithms, it is so much more difficult that we consider it to be pointless.

3.1 In-place Multiplication

Let $a, b \in R[x]$. The algorithm for in-place polynomial multiplication (IUPMUL) computes the product $c = ab$ using the Cauchy product rule

$$c_k = \sum_{\max(0,k-db) \le i \le \min(k,da)} a_i b_{k-i} \quad \text{for } k = 0..da + kb$$

where $da = deg(a)$ and $db = deg(b)$. The reason for this choice over the more familiar iteration

$$c_{i+j} = a_i b_j \quad \text{for } i = 0..da \text{ for } j = 0..db$$

is that we can sequentially write down the product with additional space for only two scratch registers. The size of the scratch registers depends on R and is given below.

Algorithm IUPMUL: In-place Univariate Polynomial Multiplication

```
IPUPMUL( (a,b,c):Array R, (da,db):Z, (t1,t2):R, m:R ): Z
 - Inputs: univariate polynomials a, b over R of degree da and db
 - working storage registers t1, t2 and space for the product in c
 - and the modulus m in R
 - Outputs: degree of the product and the product in c
     if da=-1 or db=-1 then return -1
     dc := da + db
     for k in 0 .. dc repeat
          copyinto(0_R,t1)
          for i in max(0,k-db) .. min(k,da) repeat
               InPlaceMul(a[i],b[k-i],t2)
               InPlaceAdd(t2,t1,t1)
          InPlaceRem(t1,m)
          copyinto(t1,c[k])
     - compute the degree of the product
     while dc ≥ 0 and c(dc) = 0_R repeat dc := dc - 1
     return dc
```

Note that the algorithm allows values to accumulate hence removing the remainder operation from the inner loop which often saves over half the work in practice. In the case of $\mathbf{Z}_n$ integer division is relatively expensive for small n compared with integer multiplication. In Maple, integer multiplication is about 3 times faster than integer division.

We also implemented a non-in-place version of the Karatsuba multiplication algorithm for $\mathbf{Z}_n$ for comparison. Note that the break even point will depend on the size of n as well as the degree of the polynomials. For a 20 digit prime we found the break even point to be around degree 64 indicating that IUPMUL is indeed quite efficient.

Implementation Notes We assume the following functions in R. The utility operation copyinto(x, y) copies the value pointed to by x into the space pointed to by y. The function InPlaceMul(x, y, z) computes the product xy in the space pointed to by z. Likewise InPlaceAdd(x, y, z) and InPlaceSub(x, y, z) compute the sum and difference respectively of x and y in the space pointed to by z. The function InPlaceRem(x, y) computes the remainder of x divided y in the space pointed to by x. A note about the representation.

The arrays a, b, c are arrays of pointers to pieces of storage which must be large enough to hold the largest possible values in R. When R is $\mathbf{Z}_n$ the temporaries $t1, t2$ need to be large enough to be able to accumulate at most $min(da, db) + 1$ integers of magnitude at most $(n-1)^2$. When R is $\mathbf{Z}_p[x]/(a)$ where a is a polynomial of degree $k > 0$, p is a word size prime modulus and elements of R are represented as dense arrays of coefficients, then, the temporaries $t1, t2$ need to be able to store a polynomial of degree $2k-2$ hence $2k-1$ words.

3.2 In-place Division with Remainder

In the in-place algorithm for polynomial division over R we again employ an on-line algorithm to compute the coefficients of first the quotient q then the remainder r of a divided b requiring additional space for two scratch registers. As was the case for multiplication we can remove the remainder operation from the inner loop allowing values to accumulate.

Algorithm IUPDIV: In-place Univariate Polynomial Division

```
IUPDIV( (a,b):Array R, (da,db):Z, (t1,t2):R, (lb,m):R ): Z ==
 - Inputs: univariate polynomials a, b ≠ 0 over R of degree da and db,
 - working storage t1, t2, the inverse lb of the leading coefficient of b,
 - and the modulus m in R
 - Outputs: the degree of the remainder dr where the quotient of a
 - divided b is in a[da − dq..da] and the remainder is in a[0..dr]
    if da < db then return da
    dq := da−db
    dr := db−1
    for k in da..0 by −1 repeat
        copyinto(a[k],t1)
        for j in max(0,k−dq)..min(dr,k) repeat
            InPlaceMul(b[j],a[k−j+db],t2)
            InPlaceSub(t1,t2,t1)
        InPlaceRem(t1,m)
        if t1 < 0_R then InPlaceAdd(m,t1,t1)
        if k ≥ db then
            InPlaceMul(lb,t1,t1)
            InPlaceRem(t1,m)
        copyinto(t1,a[k])
    - now compute the degree of the remainder
    while dr ≥ 0 and a[dr] = 0 repeat dr := dr − 1
    return dr
```

An additional advantage of this on-line algorithm is that if one only needs the quotient then the algorithm (modified to count from da down to the db computes the quotient without computing the remainder hence saving half the work for the case $da = 2db$.

3.3 In-place Gcd

The functionality of algorithm IUPDIV yields a simple in-place algorithm (IUPGCD) for computing Gcd's over R. Note: algorithm IUPGCD returns an unnormalized Gcd.

Algorithm IUPGCD: In-place Univariate Polynomial GCD

IUPGCD((a,b):Array R, (da,db):**Z**, (t1,t2,t3,t4):R, m:R): (Array R, **Z**)
- Inputs: univariate polynomials a and b over R of degree da and db
- additional working storage $t1, t2, t3, t4$ and modulus m in R
- Outputs: the degree of the Gcd and a or b which contains the Gcd

```
    if da < db return IUPGCD(b,a,db,da,t1,t2,t3,t4,m)
    if db = -1 return(b,db)
    while db ≥ 0 repeat
        copyinto(b[db],t3)
        copyinto(m,t1)
        InvInPlace_R(t3,t1,t2,t4) - t3 contains the inverse
        dr := IUPDIV(a,b,da,db,t1,t2,t3,m)
        (a,b) := (b,a) - interchange pointers only
        da := db
        db := dr
    return(a,da)
```

In this case additional scratch registers are needed by InvInPlace to compute the inverse (if it exists) of an element of R using the half extended Euclidean algorithm see [10] in-place. That is given a, b in R we solve $sa + tm = g$ for s. If $g = 1$ then s is the inverse of a modulo m. We have implemented the half extended Euclidean algorithm for the Euclidean domains $\mathbf{Z}$ and $\mathbf{Z}_p[x]$ where p is a word size prime in-place. Note also with slight modifications, algorithm IUPGCD can be extended to compute univariate polynomial resultants over R.

4 The modp1 Function in Maple

In this section we give further details about the Maple implementation. for arithmetic in $\mathbf{Z}_n[x]$. Note, we have not implemented the case $R = \mathbf{Z}_p[x]/(a)$ internally. Our first attempt at implementing fast arithmetic in $\mathbf{Z}_p[x]$ began with the idea that we should write the key functions (multiplication, quotient and remainder, Gcd and resultants) in the *mod* package in C. That is, the data representation for $\mathbf{Z}_n[x]$ would be a Maple general sum of products data structure. The interface would be via the *mod* function and the coding would require conversions from the Maple representation to an internal dense array representation. However, it became clear early on that the conversion overhead, was very expensive. Or, correctly put, the real work being done could be done very fast. Even for polynomials of degree 100, the time spent converting took much longer than any of multiplication, quotient and remainder, or Gcds.

The *modp*1 function in Maple does univariate polynomial arithmetic over $\mathbf{Z}_n$ using special data representations. Modp1 handles both the case of a word size modulus n separately from the case where the modulus n is large. The case of $n = 2$ is also treated specially. The actual data representation used depends on the size of n. If

$$n < \text{prevprime}\ \lfloor \sqrt{MAXINT} \rfloor$$

where MAXINT is the largest positive integer representable by the hardware, e.g. on 32 bit machines using signed integers, MAXINT $= 2^{31} - 1$, then a polynomial is represented internally as a dense array of machine integers. Classical algorithms are used with tricks to speed up various cases. For example, for the case $n = 2$ bit operations are used. For otherwise a small modulus additions in polynomial multiplication and division are allowed to accumulate if they cannot overflow. Note the prime here is used for a random number generator. If the modulus n is greater then this number, the a polynomial is represented as a dense array of pointers to Maple integers (multi-precision integers). And the in-place algorithms described in the previous section are used. A example of usage for a large modulus follows

```
> p := prevprime(10^10);
                              p := 9999999967

> a := modp1( Randpoly(4), p );

         a := [1110694326, 3633074819, 4256145114, 8458720791, 7419670467]

# This represents the polynomial
> modp1( ConvertOut(a,x), p );
                    4                3                2
    7419670467 x  + 8458720791 x  + 4256145114 x  + 3633074819 x + 1110694326

> b := modp1( Randpoly(4), p );

         b := [2062222184, 2974124144, 4305615901, 5580039851, 6753832980]

> g := modp1( Randprime(4), p );

             g := [4685305298, 2712797428, 1717237881, 3687530853, 1]

> ag := modp1( Multiply(a,g), p ):
> bg := modp1( Multiply(b,g), p ):
> modp1( Gcd(ag,bg), p );

                [4685305298, 2712797428, 1717237881, 3687530853, 1]

> modp1( Factors(ag), p );

 [7419670467, [[[3203615647, 1], 1], [[7211058641, 1284247953, 9477941733, 1], 1],

                  [[4685305298, 2712797428, 1717237881, 3687530853, 1], 1]] ]
```

Where note the output format of the Factors function is

$$[u, [[f_1, e_1], \ldots [f_n, e_n]]] = u \times f_1^{e_1} \times \ldots \times f_n^{e_n}$$

Note, the **mod** function in Maple provides a nicer interface for the user to these facilities. For example, in the following, the Maple polynomials will be automatically converted into the *modp1* representation where the computation is done, then converted back on output. The *evalgf1* package uses the *modp1* facility to implement efficient univariate polynomial arithmetic over finite rings and fields $\mathbf{Z}_p[x]/(a)$. Polynomials are represented as dense arrays of *modp1* polynomials. This facility is also accessed via the `mod` function with conversions taking place automatically. For example

```
> f := x^8+x^4+x^3+x+1;
                                    8    4    3
                            f := x  + x  + x  + x + 1

> Factor(f) mod 2;
                                 8    4    3
                                x  + x  + x  + x + 1

> alias(a=RootOf(f,x)):
# Factor f over GF(2^8) = Z2[x]/(f)
> Factor(f,a) mod 2;

      6    3    2              6    4    3    2                        7    6    5    2
(x + a  + a  + a  + 1) (x + a  + a  + a  + a  + a) (x + a) (x + a  + a  + a  + a )

         2         4         4    3                     7    6    5    4    3
   (x + a ) (x + a ) (x + a  + a  + a + 1) (x + a  + a  + a  + a  + a  + a)

> Expand(") mod 2;
                                 8    4    3
                                x  + x  + x  + x + 1
```

We make some comments about the efficiency gain compared with Maple 4.2 where most of these operations were interpreted. Polynomial multiplication over $\mathbf{Z}_n$ is about 15-30 times faster for a small modulus and polynomial quotient and remainder, Gcd and resultant, and factorization, are all $100-400$ times faster. For the case of a large modulus, the improvement is typically a factor of $3-15$ where the larger the modulus, the less the improvement. We will not present here any timing comparisons with other systems. We refer the reader to [20] for timing comparisons for polynomial factorizations over $\mathbf{Z}_p$ and $\mathbf{Z}$. We do mention that Maple V can compute the resultant in SIGSAM problem #7 [11] in 26 seconds on a Vax 11/85 and factor it in a further 26 seconds.

We list here a summary of which functions in the *modp1* package we have coded in C and which we have coded in Maple. Note this depends on whether the modulus is small or large. And that all operations coded in C run in linear space.

Operation	Small	Large	Description
Degree	Y	Y	
Ldegree	N	N	The degree of the first non-zero term
Coeff	Y	Y	The coefficient of x^i
Diff	Y	N	The derivative
Shift	Y	Y	Shift a polynomial by x^n where $n \in \mathbf{Z}$
Add	Y	Y	+
Sub	Y	Y	−
Multiply	Y	Y	×
Power	N	N	^
Rem	Y	Y	Polynomial remainder (optionally computes the quotient)
Quo	Y	Y	Polynomial quotient (optionally computes the remainder)
Divide	N	N	Polynomial exact division
Gcd	Y	Y	Polynomial greatest common divisor
Resultant	Y	N	Polynomial resultant
Gcdex	Y	N	The extended Euclidean algorithm
Eval	Y	Y	Polynomial evaluation
Interp	Y	N	Polynomial interpolation
Powmod	Y	N	Compute Rem(a^n,b) using binary powering
Randpoly	Y	N	Generate a polynomial with random coefficients
Sqrfree	N	N	Square-free factorization
Roots	N	N	Compute the roots of a polynomial
Irreduc	N	N	Irreducibility test
Factors	N	N	Factorization (Cantor-Zassenhaus distinct degree)

5 Conclusion

We found that by implementing the primitive operations addition and subtraction, multiplication, quotient and remainder, Gcd and resultant, evaluation and interpolation in $\mathbf{Z}_n[x]$ in C, one is able to then write efficient code for factorization over $\mathbf{Z}_p$ and polynomial resultants, Gcds, and factorization over $\mathbf{Z}$ in a high level language. Even if the high level language is interpreted, the efficiency lost is negligible because the bulk of the work done is the primitive operations. This investment in systems code can also be used to implement several other important algorithms efficiently. Firstly, Collins modular method for Gcds and resultants [9] of dense multivariate polynomials over $\mathbf{Z}$ which we have implemented for bivariate polynomials in Maple. Secondly, efficient polynomial arithmetic over finite fields and rings given by $\mathbf{Z}_p[x]/(a)\ :\ a \in \mathbf{Z}_p[x]$ including resultants, Gcds and factorization. Thirdly, efficient univariate polynomial Gcd computation over a simple algebraic extension of $\mathbf{Q}$, [17, 18, 14].

We also found that the careful use of in-place arithmetic for the key operations multiplication, quotient and remainder over $\mathbf{Z}_n$ results in a significant overall efficiency gain. The gains made by using in-place arithmetic are first that we are able to essentially eliminate the overhead of storage management. Second, we save operations by allowing values to accumulate temporarily. Our implementation in Maple typically results in improvements of factors of 3 to 5 depending on the size of the modulus n.

If computer algebra systems are to get the most efficiency out of the systems hardware for basic arithmetic domains $\mathbf{Z}$, $\mathbf{Z}_n$ and $\mathbf{Z}_p[x]/(a)$, then we believe that this will happen only given careful attention to the functionality provided so that we can build polynomial, vector and matrix arithmetic over these domains without interaction with storage management at every operation. This issue of storage management is even more acute in parallel systems.

Finally, what about linear algebra over $\mathbf{Z}_n$? Although we have not considered vector and matrix operations, it is not difficult to imagine similar schemes whereby one is able to perform vector and matrix arithmetic, such as determinants, in-place. We expect that one would see comparable improvements in performance. Thus it would seem that the next thing to do is to implement a similar facility to the *modp1* facility for vector and linear algebra where the key routines will be matrix multiplication and Gaussian elimination.

References

1. Berlekamp E.R. Factoring Polynomials over Finite Fields. *Bell System Technical Journal,* No **46** 1853-1859, 1967.
2. Berlekamp E.R. Factoring Polynomials over Large Finite Fields. *Mathematics of Computation,* **24** 713-715, 1970.
3. Brown W.S. On Euclid's Algorithm and the Computation of Polynomial Greatest Common Divisors. *JACM,* **18**, 478-504, 1971.
4. Brown W.S., Traub J.F. On Euclid's Algorithm and the Theory of Subresultants. *JACM,* **18**, 505-514, 1971.
5. B. Buchberger, A Theoretical Basis for the Reduction of Polynomials to Canonical Forms, ACM SIGSAM Bulletin, **9**, (4), November 1976.
6. Cantor D.G., Zassenhaus H. A New Algorithm for Factoring Polynomials over a Finite Field. *Mathematics of Computation,* **36**, 587-592, 1981.
7. Char, B.W., Geddes K.O., Gonnet, G.H. GCDHEU: Heuristic Polynomial GCD Algorithm Based On Integer GCD Computation. *Proceedings of Eurosam 84*. Springer-Verlag Lecture Notes in Computer Science, **174**, 285-296, 1984.
8. Collins G.E. Subresultants and Reduced Polynomial Remainder Sequences. *JACM,* **14**, 128-142, 1967.
9. Collins G.E. The Calculation of Multivariate Polynomial Resultants. *JACM,* **18**, 515-532, 1971.
10. Geddes K.O., Labahn G., Czapor S.R. *Algorithms for Computer Algebra.* To appear 1992.
11. Johnson S.C., Graham R.L. Problem #7 SIGSAM Bulletin issue number 29, **8** (1), February 1974.
12. Kaltofen E. Computing with Polynomials Given by Straight-Line Programs: I Greatest Common Divisors. *Proceedings of the 17th Annual ACM Symposium on the Theory of Computing,* 131-142, 1985.
13. Knuth, D.E. *The Art of Computer Programming* Vol. 2: *Seminumerical Algorithms* (Second Edition). Addison-Wesley, Reading Massachusetts, 1981.
14. Langemyr, L., McCallum, S. The Computation of Polynomial Greatest Common Divisors. *Proceedings of EUROCAL,* 1987.
15. Maple V Language Reference Manual Springer-Verlag, 1991.

16. R. Peikert, D. Wuertz, M. Monagan, C. de Groot Packing Circles in a Square: A Review and New Results. IPS Research Report No. 91-17, September 1991 ETH-Zentrum, CH-8092 Zurich, Switzerland.
17. Geddes K.O., Gonnet G.H., Smedley T.J. Heuristic Methods for Operations with Algebraic Numbers. *Proceedings of the ACM-SIGSAM 1988 International Symposium on Symbolic and Algebraic Computation*, ISSAC '88, 1988.
18. Smedley T.J. A New Modular Algorithm for Computation of Algebraic Number Polynomial Gcds. *Proceedings of the ACM-SIGSAM 1988 International Symposium on Symbolic and Algebraic Computation*, ISSAC '89, 91-94, 1989.
19. David Yun, The Hensel Lemma in Algebraic Manipulation. Ph.D. Thesis, Massachusetts Institute of Technology.
20. Paul Zimmerman, A comparison of Maple V, Mathematica 1.2 and Macsyma 3.09 on a Sun 3/60. mathPAD newsletter **1** (2), April 1991. Gruppe mathPAD, Universität Paderborn.
21. Zippel R. Probabilistic Algorithms for Sparse Polynomials. *Proceedings of Eurosam 79.* Springer-Verlag Lecture Notes in Computer Science, No 72, 216-226, 1979.

LILK - Static Analysis of REDUCE Code

B D Brunswick

University of Cambridge Computer Laboratory
New Museums Site, Cambridge
Brian.Brunswick@cl.cam.ac.uk

1 Abstract

One of the advantages of fancy compile time type checking systems besides avoiding runtime checking overhead is the early detection of program bugs and conceptual problems. Unfortunately, large bodies of old code exist which were written for untyped or dynamically typed systems, which leave such problems to be detected during execution, often much later and only in response to peculiar or unusual input. A typical example of this syndrome is the computer algebra system REDUCE. The experience of fighting valiantly with this in the past, has led now to an attempt to improve on the situation by applying inferencing techniques to add type information to the current source code in a relatively painless way. I present here a preliminary report on this work, showing that the complexly related type scheme of an algebra system is particularly suitable for this.

2 Introduction

REDUCE the computer algebra system is implemented in RLISP. As its name implies, this is a derivative of LISP, and consists of a parser for an Algol-style syntax, and some syntactic sugar for various iterating constructs that are obscurely named in LISP. It is based on something called Standard LISP, which is an extremely cut down core language that can easily run on smaller machines. RLISP is also somewhat enhanced and presented to the user as the control language for REDUCE. In this guise it has some minimal typing information embedded to do with distinguishing algebraic and symbolic computation and scalar and matrix quantities, but the implementation language lacks even these. The only type system that is imposed is the underlying minimalist dynamic typing of a LISP core, where the main distinction is between atoms and cons pairs, which make up lists.

To capture the primary range of types used in algebra the REDUCE system builds more complex types up from nested cons pairs. For instance, a term uses two pairs to tie together the coefficient, variable and index. A polynomial is a list of terms, and a standard quotient is a pair of polynomials. This standard quotient is the main object with which REDUCE computes, and its structure is further complicated when dealing with multivariate polynomials or polynomials in things other than simple variables, where both the coefficient and variable entries in a term will themselves be lists.

Many parts of REDUCE involve working with smaller parts of this structure, and the problem is that there is no means for the LISP type checking to distinguish one type from another, until a decomposing operation fails at some point possibly deep in the code, and far distant from the point where a type error occurred. The feeling is that since REDUCE as an algebra system deals with a relatively small range of closely related types, it should be possible to take the existing code and add type information to it, sufficient to allow compile time type checking for these sort of errors, and also to allow faster, non-checking versions of the LISP primitives to be used.

So the motive behind the work has been to automatically apply the sort of type inferencing techniques used in newer languages to this substantial body of old code with the hope of improving documentation and maintainability of the code, finding type-style bugs which code layered on LISP is particularly prone to, and incidentally providing an intelligent formatting tool. It has turned out that a usable browser for code has become a desirable goal too, from its usefulness in aiding the comprehension of code.

Another welcome possibility is that some of the more delicate and complex algorithms present in REDUCE may be liberated into a suitable form for re-use and similarly to improve the modularity and independence of the REDUCE code itself.

3 The Approach

I have elected to approach this by providing a method by which (partially annotated) code may be read into the system and then processed in some way, such that it can still be reproduced as (fully annotated) text without losing the human imposed structure and extra information that conveys vital parts of its readability. Of course, morally, this is somewhat the wrong solution to the dilemma of editing and producing code. Using text editors to work on something that is fundamentally much more than just a sequence of characters has always been a poor compromise, and its inadvisability is only further exemplified by profiling tests on compilers that demonstrate that up to half their time is spent parsing the input. It does, though, sidestep all the issues of human interface and utility design that I am not primarily interested in, and enable one to apply analysis and annotation techniques.

This does introduce the additional problems of parsing, storing and reproducing the additional information that is normally discarded on input of a language – namely comments and formatting information. At present, the latter is mostly discarded in favour of simply generating anew a nice layout on output, but the scarcity of comments in some portions of the code makes them only more valuable to retain.

The actual checks that can be applied to the code are very much similar to those done by a good quality compiler or lint like tool. This means basically type checking on both a local and global level (including scope considerations) and data flow investigations. The latter has really no real interest as the dynamically typed nature of the world has no influence on it, and the standard algorithms should be used. It is a pity that it seems that most LISP compilers do not do this – it should be easy to add, and become as common as it is in other languages.

4 Overview of Problems

Several problems have arisen along the way. The most irritating by far in that it potentially forces actual rewriting of code for little purpose is where locals are used with varying types throughout a procedure. In the dynamically typed-come-untyped LISP environment, this is perfectly legal, and natural to an extent, say when a given value is being refined through the type scheme, or for trivial temporary variables. Such situations are not indicative of program logic errors. They can be dealt with in most cases by the use of dataflow and live variable analysis, which distinguishes separate use of the same variable. Of course, such analysis is a desirable thing in itself in that it can detect dataflow anomalies in a useful way, so there is some gain.

Another problem is that of how to present and react to type conflicts when they occur. This situation is expected to be moderately common during analysis, and so in contrast to other type inferencing systems there is need to be interactive. This means that the theoretical notion of decideability which has so dominated some investigations is now replaced by practicality. It is important for utility that when whatever action is decided upon is taken, it be possible to backtrack and reuse most of the work that has been done so far, along with suitable tracking of the (valuable) users decisions in case of accidents to the system. In order to present the user with sensible information, it is important to have some kind of accountability on type decisions, so that it is possible to quickly get to the root cause of the problem. This kind of detailed recording of calculation leads both to the possible problem of data accumulation, and also, when such things are presented to the user to a rapid bloating of output if all available data is to be produced. Interactivity really is the only way to go in this kind of exploratory investigation of code structure and large numbers of relations, and using a browser the only convenient way to provide access. (One irritating point is that the interaction has been restricted by being though character based devices, and the point and click nature of a mouse is sorely missed. I am sure that even a minimal movement in this direction would be useful, and intend to implement a minimal such interface on the platform that I use as part of further work.) What has in fact been produced so far is an ability to change the code of one procedure at a time, and then remove from the database all types deduced as a consequence of it, and restart from there.

5 Pretty Reading

There has been the choice in the design of the code reading and output system of whether to go for exact reproduction or an approximation. Since the goal is to add annotation that is likely to force changes in the layout, it has seemed natural to adopt more of the techniques of a beautifier program, since that also gives that functionality as a free bonus. The sort of techniques involved here are fairly simple, with the only somewhat delicate part being the handling of comments. The usual seems to be to distinguish between comments alone on a line, which often form blocks and may well be specially formatted in a way that should not be disturbed, and those postfixed to lines, which are relevant to that particular line. These latter can reasonably be regarded as straight text, and possibly split or combined if the formatting of the program changes.

Given that I have this classification, the further problem arises in this case of how the comments are to be stored internally, and what particular program structure they are to be associated with. I have decided that block comments are to be associated either with the following defined object if encountered at the outermost level, or with the next object that occurs in a sequence. If a block occurs that is not at a natural break in some sequence, then its contents is regarded along with any postfixed comments as applying non-locally to the current expression. The syntactic extent of the expression corresponding to particular comments is clearly something not easily precisely defined in the context of RLISP which does not have a syntax that always distinguishes statements. I have chosen to regard it being as at least as much as occurs on one line, combining lines except in the case where the line itself forms a clean subexpression, possibly with an operator adjoined. (Given a known set of operators and delimiters)

So the extraneous data that must be kept along with forms in memory is either a set of lines of text corresponding to the comments found before an object definition, or that for some subexpression, which wants to be kept associated with that expression. This is fairly naturally done by using a wrapper around the expression, placing it where the expression would otherwise be. This is entirely satisfactory for most expressions, as it means that the result can still be executed provided the wrapper is suitably defined, but caution is necessary in situations where the expression is a special form. I avoid this by redefining the forms involved to check for the situation. (It is not actually clear that retaining executability is necessarily appropriate, but it seems more convenient to retain if if possible – this is one issue that may be readdressed.)

6 Type Annotation

The first issue of type annotation is its internal/external representation. There must be settled several contributory points, namely a type representation schema, which is discussed below, but also the point of the precise extent of such information as transmitted to the user. Internally, there is a type annotated to each subexpression, and also a type associated with each live variable at a given point in code. This type may not necessarily be fixed for a given variable, but the normal, preferable situation is that it is, when it can be treated as a conventional type declaration. A distinction must be made between variables initialised to NIL as an initial value, and those for which NIL is effectively an undefined value, illegal for its type. In a corrected situation, these will be distinguished by the liveness after its definition, but some representation of this is also desirable for those to whom liveness is not implicitly immediately apparent! I opted for output of explicit initialisation in those cases where it is live, in an extension to the syntax. With the less preferable state of differing types, some flag of this fact should be at the variables definition, and the expression type annotation of the assignment be sufficient clue to the programmer. This fits in reasonably with the notion that the extent of expression annotation should be configurable. As a simple scheme, output is done only for expressions with top level using one of those operators which are flagged. Initially assignment, and special forms corresponding to blocking are selected.

The task of type annotating the program is necessarily a somewhat delicate one. The whole point is to detect mistakes so it is expected to find errors, and one must be prepared to try to ease the job of correcting or dismissing them. For this purpose it is certainly important to use a breadth first style scan of the available inferences, so that conflicts are detected with as much avoidance of long chains of stored reasons as possible. This also means smaller amounts of backtracking as well. In order to perform these operations of backward reasoning and forward removal of consequences, bidirectional links are required, and these might be thought to represent a significant cost in terms of memory. Fortunately, the actual number of objects being typed for which information is retained permanently is relatively small compared to the amount of program code involved.

The internal representation of types turns out not to be as important as might be thought. This is because the actual process of inferencing does not actually use them other than as the already known types of functions. Of course, when typing structure decomposing and constructing functions the representation is important, especially in the context of REDUCE embedded in LISP, where these functions are highly variadic. However, part of what I am setting out to do is precisely to eliminate and/or characterise such usage, and convert it to specific selector functions, and so do not want to embed issues of data respresentation into the typed output. The typing of functions is itself part of it of course, leading back to the question of representation, but the situation is simplified by the fact that there is very little dealing with higher order functions acting on other functions. I have chosen to represent program data types with simple atomic symbols. Functions will be represented by a simple disjunction of type mappings, from a number of argument types to one result type, corresponding with the restrictions of the standard-LISP model, which REDUCE assumes. These disjunctions, will, however only be created at the users command, since they could represent precisely the kind of type conflict that one is trying to find.

The actual algorithm used for type inferencing and checking is fairly simple. Each expression in a procedure gives rise to a type equation, with extra unknown type variables added for intermediate values. There is no attempt at a partial unification system however, and all types used are completely defined. Possible inferences from a procedure go onto a queue in the order that they are encountered while examining the expression, and then those that have enough information defined to identify the type amongst possible alternates for the function are used to further spread the types. Functions whose type is unknown can only be used when all arguments are typed. Once everything possible is done within one procedure, the intermediate variables are discarded to avoid the buildup of unnecessary information, and only the completed types for variables (including dataflow) and functions are retained, possibly with the flag indicating that not all was able to be deduced about that procedure and that it should be returned to. Procedures will be examined eventually in an order determined by a complexity measure, but presently just as nominated by the user.

Whenever a type is assigned to a function or variable (things that will be kept) there are two links made for the purpose of reason chaining and undo. The undo link is not actually presently used, as a simple journal of symbols operated on is used instead. This also provides a method for checkpointing the progress so far, since information within procedures apart from locals is not kept, and the possible loss of more than is necessary has not become a problem while small scale experimentation is going on.

7 Results

At the time of writing I do not yet have any statistics on anomalies and number of interventions needed over a large area of program code, so results are confined to impressions received while using the program.

Perhaps the clearest conclusion so far is that it is a delicate task to control the amount of output generated and present it in a comprehensible way to enable tracing of the cause of type conflicts. Overwhelmingly long, complex chains of inferences are easily produced, and seem to point the way towards needing better interaction techniques.

Dataflow analysis does solve most problems involving multiple use of local variables, and surprisingly few manual changes are needed in this direction. This is a good situation, but I do not yet have a really satisfactory output syntax for representing what happens.

Instances of actually finding bugs have been disappointingly rare. (To this project, not users of REDUCE!) This is possibly due to the restriction up to now to attempts on relatively easy sections of code, where providing sufficient seeds from which to start the inferencing is less time consuming. The choice of these seeds while not crucial does have a bearing on the inferencing, and choosing the types needs an occasionally non-trivial understanding of the code. This leads back to the utility of the simple browser in increasing this comprehension.

8 Examples

I here present a very tiny sample of output produced. The types involved are *sq*, short for standard quotient, i.e. fraction, and *form*, a polynomial which forms the numerator and denominator of the fraction. There are selector and constructor functions whose type has been provided, and other types are inferred by the system. First there is the original source code of the function *multsq*, followed by output before and after having scanned it, and the formatted but minimally annotated version. It should be noted that the system understands the boolean functions, but does not list them out, and treats equality and *null* specially. The types are supplied with a recogniser function for their constants, which recognise 1 and *nil* as valid *forms*.

```
symbolic procedure multsq(u,v);
   % U and V are standard quotients.
   % Value is canonical product of U and V.
   if null numr u or null numr v then nil ./ 1
    else if denr u=1 and denr v=1 then multf(numr u,numr v) ./ 1
    else begin scalar x,y,z;
        x := gcdf(numr u,denr v);
```

```
        y := gcdf(numr v,denr u);
        z := multf(quotf(numr u,x),quotf(numr v,y));
        x := multf(quotf(denr u,y),quotf(denr v,x));
        return canonsq(z ./ x)
    end;
```

...

```
*** List of known types

smacro procedure numr u : sq : form;          % Numerator of standard
                                              % quotient.

smacro procedure denr u : sq : form;          % Denominator of standard
                                              % quotient.

smacro procedure u : form ./ v : form : sq; % Constructor for standard
                                              % quotient.

*** Processing multsq

*** List of known types

smacro procedure numr u : sq : form;          % Numerator of standard
                                              % quotient.

smacro procedure denr u : sq : form;          % Denominator of standard
                                              % quotient.

smacro procedure u : form ./ v : form : sq; % Constructor for standard
                                              % quotient.

symbolic procedure multf( u : form, v : form ) : form;

symbolic procedure gcdf( u : form, v : form ) : form;

symbolic procedure quotf( u : form, v : form ) : form;

symbolic procedure canonsq( x : sq ) : sq;

symbolic procedure multsq( u : sq, v : sq ) : sq;
```

...

```
symbolic procedure multsq( u : sq, v : sq ) : sq;
% U and V are standard quotients.
% Value is canonical product of U and V.
if null numr u or null numr v then
  nil ./ 1
else
  if denr u = 1 and denr v = 1 then
    multf( numr u, numr v ) ./ 1
  else
    begin
      scalar z : form;
      scalar y : form;
      scalar x : form;
      x := gcdf( numr u, denr v );
      y := gcdf( numr v, denr u );
      z := multf( quotf( numr u, x ), quotf( numr v, y ) );
      x := multf( quotf( denr u, y ), quotf( denr v, x ) );
      return canonsq( z ./ x )
    end;
```

9 Conclusions

While final conclusions are obviously not yet possible, it is still possible to reach some clear points. Firstly, this technique is possible and useful, in that it can reveal bugs and make code clearer, and correspondingly easier to maintain and upgrade. On the other hand, without very great care the quantity of output can obscure the real point underlying some possible problem with code, and tracking down and repairing such problems is a job much akin to bug fixing. (Well it is!) The trace information from type inferencing helps with this quite significantly however, along with the bones of a browser for the code, and could be much improved to provide a useful tool.

10 Future Aims

There are several clear directions from which to go with this project. Obviously, a more complete set of statistics and results are desirable, together with improvements with the formatted output to enable a clean and useful final result. To enable this to be done with pleasure, I will improve the user interface to enable easy editing and browsing, and possibly incorporate some mouse support, though this will be of necessity fairly limited in that I do not wish to spoil the portability that is one of REDUCEs main benefits.

The degree of speed gain attainable by allowing the compiler to generate unchecked primitive operations (where type information indicates) needs to be investigated, as this is possibly a valuable gain.

Another desire is to refine the type system to incorporate more of the delicate distinctions that are present in an algebra system, such as simplified quotients and monic polynomials. These are often difficult to capture using the model of (multiple) inheritance used in some languages, but should be possible here.

11 References

Anthony C. Hearn: The REDUCE User's Manual, 1991, RAND Publication CP78
UNIX Programmer's Manual
Any works on type inferencing systems

The Extensions of the Sisyphe Computer Algebra System: ulysse and athena

C. Faure[1], A. Galligo[1], J. Grimm[2], L. Pottier[2]

[1] INRIA and Université de Nice-Sophia Antipolis: Laboratoire de Mathématiques 06108 Nice CEDEX, FRANCE

[2] Institut National de Recherche en Informatique et Automatique 2004 route des Lucioles, Sophia Antipolis 06565 Valbonne CEDEX FRANCE

1 Introduction

In [GGP90] we presented the main characteristics of SISYPHE, a prototype computer algebra system, written in LELISP, developed since 1988 in the SAFIR project, at INRIA Sophia Antipolis and Université de Nice.

There, we concentrated on 4 aspects:

- compilation and modularity, reasonably small kernel.
- algebraic structures easy to build, and efficient generic algorithms.
- formal simplifier using rewriting techniques.
- extensible graphic user interface [Kaj90, Kaj91].

The development and integration of these features into a single easy to use system have now led us to construct a new architecture.

The main innovations are the creation of a separate parametrized formal simplifier named **ulysse** (also written in LELISP [INR89] like SISYPHE), as well as an extended algebraic type system **athena**. So, the system consists of integrated autonomous components which provide :

- Powerful formal simplifications.
- Types for mathematical structures.
- Generic algorithmic library.

In the first three sections of this paper we will describe **ulysse** and its standard type system, then the extended algebraic type system **athena**. In §4 we present the new I/O and programming language of SISYPHE, and we briefly describe the current algebraic library. Finally in §5 we give some examples of the flexibility of our approach.

2 The ulysse System

Following the example of Scratchpad [DGJ+84] for mathematical algebraic structures, we chose to separate formal objects used in simplification of mathematical expressions from algorithms.

So the only things hidden behind algorithms are algorithms ....

To express the notions *mathematical structures, formal simplifications and algorithms* we chose to use the classical Tarski semantics for predicate logic and its restriction to *universal algebra*, in which these three concepts become *signatures, equational theories and models or Σ-algebras.*

Signatures give the syntax of formal expressions and define *abstract algebras* (whose elements are *terms*), equations express formal properties on terms (the operational aspect being *rewrite rules*), and models are sets with functions, i.e. data structures with algorithms.

For example, the additive group of integers in **ulysse** is given by its signature $\{+, 0, -\}$, its axioms $\{(x+y)+z = x+(y+z), x+y = y+x, x+0 = x, x+(-x) = 0\}$, and its data structure in LeLisp (bignum, functions *add*, *zero*, *opp*).

The **ulysse** system has then two major parts :

The basis which contains the signatures (operators with functional types), the syntax for operators (the parser and the printer), the equational properties and rewriting rules attached to the operators, and the *interpretations* of the operators, which are algorithms of the models.

The engine which uses the informations in the basis to treat expressions given at the top-level, following a given strategy (for example "simplify with equations, rewrite with rewrite rules, and evaluate with interpretations")

Note that with this organization, the main objects are *operators*, unlike Scratchpad for example, in which the main objects are *domains*. The motivation for this "operator-oriented" approach is that we focus in **ulysse** on syntactic operations on mathematical expressions. The semantic part is contained in the notion of models (= domains), and is developed by the two main tools available from **ulysse** : **athena** (mathematical types, and automatic generation of algorithms) and SISYPHE (algorithms and programming language).

Let us detail now the two components of **ulysse**:

2.1 The Basis

It has three components, consisting of only *declarative* information :

Properties of operators, normal forms are of two types. The first one is syntactic and concerns the parsing and the printing of terms (which are made by a parameterised extension of those of SISYPHE.) The second type of property is semantic and consists of sets of equations, chosen from a library of about 15 equation schemes (e.g. associativity, morphisms, homogeneity, ...).

Rewrite rules express more complex properties of operators. Variables can be typed, to constrain rule applications. Rewriting is allowed modulo associativity and/or commutativity of operators.

Models and interpretations of operators which use a simple standard type system, which is close to the object-oriented types of LeLisp. A model is a type, given by a symbol and by *methods* for ordering and printing its elements. Interpretations (i.e. instantiation) of operators are given by two components: a functional type involving the previous types, and a LeLisp function, i.e. an algorithm. For example, here is an interpretation of "+" in the polynomials of SISYPHE:

$$((pol, pol) \rightarrow pol,\ pol+)$$

These types, named "numerical types" are grouped in "formal types" which represent different implementations of the same structure : conversions between two types in the same formal type are always possible. Interpretations may be defined with these formal types.
Finally, note that overloading is obtained by giving different interpretations to the same operator (even with several arities for example).

2.2 The Engine

It constitutes the *operational* aspect of the basis, and is used at the top-level of `ulysse` to transform input terms. How does it work with the different kinds of information in the basis?

Syntactic properties are used for parsing and printing.

Equations are used to transform terms into *canonical* or *normal* forms. Efficient algorithms are defined for every set of equations, using two kinds of implementation of terms (operator + list of arguments or multi-set of arguments for associative and/or commutative terms).

Rewrite rules are grouped into compiled systems of rules (i.e. a compiled LeLisp function computing the normal form of a term). A Knuth-Bendix algorithm is available in `ulysse` to complete interactively a given system of rules.

Interpretations are used *via* type unification between their functional types and types of arguments of an operator. This is done also by using conversions between types, determining the "best" interpretation when choices occur.

The strategy used by the engine to combine these transformations of terms is free and easy to define. For example we can use evaluation (= use of interpretations) before rewriting, transform sub-terms before root term, ... The standard strategy being : compute normal forms, rewrite, and evaluate, all of this done on sub-terms before root term.

3 The Algebraic Types System : athena

The types defined in **athena** are based on the classical semantic of the predicate logic, where we manipulate *terms* interpreted in *models*.

Models are sets with algorithms, typing their elements (like domains of Scratchpad). The type of a model is a *signature*, i.e. a set of functional symbols, and the type of signature is a *scheme of signatures*, giving generic functional types for elements of signatures (like categories of Scratchpad) and logical axioms satisfied by them (which justify the term "model" against the term "domain").

How are these notions implemented?

Models A model is given with its signature and its algorithms (in LELISP or other language), using eventually other models (e.g.vector space on a field). Its main characteristics are:

- a name (an arbitrary term).
- an operational description of the implementation of its elements (e.g. sparse representation for polynomials, etc).
- a reader and a printer which transform a term (e.g. $5 * X^4 + 9 * X - 7$) into internal representation (e.g. ((5 . 4)(9 . 1)(-7 . 0)) in LELISP), and the converse.
- auxiliary models.
- eventually an enumerating algorithm.

For example, here is the model of integers :

```
(1) describe('Z);

                    [              Model              ]
                    [       signature = F_Ring        ]
                    [           Algorithms:           ]
                    [         = = ("=" : Lisp)        ]
                    [         + = ("+" : Lisp)        ]
                    [         * = ("*" : Lisp)        ]
                    [compare = ("#:ul:alg:Z-order" : Lisp)]
                    [   prin = ("identity" : Lisp)    ]
                    [ read = ("#:ul:alg:read_Z" : Lisp) ]
                    [         - = ("-" : Lisp)        ]
                    [         0 = ("0" : Lisp)        ]
                    [         1 = ("1" : Lisp)        ]
```

Signatures and Schemes An example is sufficiently clear :

```
(2) describe('F_Ring);
                    [        Signature        ]
                    [       type = ring       ]
                    [      name = F_Ring      ]
                    [    read_el = read       ]
                    [    prin_el = prin       ]
                    [       equal = =         ]
                    [    order = compare      ]
                    [        law = +          ]
                    [       unit = 0          ]
                    [        sym = -          ]
                    [        mul = *          ]
                    [     unit_mul = 1        ]
                    [inherit = F_Additive_Group]

(3) describe('ring);
```

```
[                              Scheme                              ]
[                           name = ring                            ]
[                            Operators:                            ]
[equal =   [# ALG , # ALG] --> 'Boolean                            ]
[                       ftype = s_signature                        ]
[                       inherit = 'group                           ]
[   law =   [# ALG , # ALG] -->  # ALG                             ]
[   mul =   [# ALG , # ALG] -->  # ALG                             ]
[  order =   [# ALG , # ALG] --> 'Z_3                              ]
[      prin_el =   [# ALG] -->  # X                                ]
[      read_el =   [# X] -->  # ALG                                ]
[       sym =   [# ALG] -->  # ALG                                 ]
[                          type = SCHEME                           ]
[          unit =   [] -->  # ALG                                  ]
[        unit_mul =   [] -->  # ALG                                ]
[            parameters =  Ring_axioms(# ALG )                     ]
```

The `# ALG` can be compared with the $ of Scratchpad. Note that signatures and schemes support inheritance.

This type system allows to specify generic algorithms, e.g. the operator giving the power of an element of a field will have the type:

```
[#X:(#Y:field), Z] --> #X
```

where `#X` and `#Y` are type variables.

4 Algorithmic Libraries, Programming Language

4.1 Sisyphe

SISYPHE is a computer algebra system, which comes with a library, an I/O mechanism and an evaluator. Only the library is used by **ulysse**. The I/O mechanism can read Lisp files, or SISYPHE files, and compile them. It can also write objects according to different formats (general format, SISYPHE, Fortran, or LaTeX format) and is very simple to use. For instance, in TeX mode, the result of evaluating

```
format("Example: The sum of ~A and ~A is ~A.", x, y ,x+y);
```

is

```
$$
{\hbox{\rm{}Example: The sum of $\displaystyle x$ and $\displaystyle y$ %
is $\displaystyle x + y$.}}\eqno(8)
$$
```

which is interpreted by TeX as

$$\text{Example: The sum of } x \text{ and } y \text{ is } x + y. \qquad (8)$$

Complicated algorithms in the library, such as factorization and integration use the format function to output intermediate results in debug mode. This is sometimes useful, for instance to explain in formal integration why a certain expression has no primitive.

Objects of Sisyphe Objects in SISYPHE are generally Lisp vectors, containing a type and some piece of data, such are numbers, functions, arrays, and expressions. However, they are objects that do not contain enough information to be handled alone: we called them rats (for rational functions and polynomials). To use them, another object is needed, the ring in which they live. A ring is essentially a data structure containing some constants (the variables of the polynomial, the ring of coefficients, the internal representations of zero and one, etc) and functions used for computations. These functions can be as easy as addition or as complicated as factorization. Finally, a ring also contains information about representation of objects, indicating for instance ordering of monomials of multivariate polynomials.

A user can create any ring he wants, and convert any object to any ring (provided that this is possible). He can also control the ring in which the result of a given computation will be, for instance, if adding two polynomials in $\mathbf{Z}[x, y]$ yields a result independent of x, the system can automatically convert it into $\mathbf{Z}[y]$.

Algorithms Many algorithms are implemented in SISYPHE. Most of them are written in Lisp. The usual mechanism for an operation on polynomials is the following: there is a function with a simple name, say *rat+* that takes two polynomials (in internal representation) and a ring; this function looks in the ring to find the function that depends on the structure. For instance, this function may be *rat-num1-add-hs* to add two sparse univariate polynomials. This function calls the function *num-add-num-1s* by replacing the ring by its coefficient ring. Finally this function simply adds the coefficients using the generic function *rat+*.

Of course, this works only for simple functions. In the case of factorisation for instance, the object is first converted to a ring $\mathbf{Z}[x, y]$ (if not a polynomial, numerator and denominator are factored separately, common factors are cancelled by the simplifier). In the current version, factorisation is only written for polynomials with integer coefficients. In some cases, conversions are not automatic. For instance, computation of a gcd uses, as a default, a heuristic method that needs the ring of integers to be the ground ring. If this ring is the ring of rational numbers, a common denominator can be computed to have only integer coefficients. If the ground ring is the ring of integers modulo N, subresultants are used to compute the gcd, and no conversion occurs.

Programming Language SISYPHE comes with a programming language, that is also used to evaluate things at the toplevel. In fact, there is no interpreter, everything is compiled and the code is immediately executed. Let us see some features of this language.

First of all, evaluation of variables is handled as in general languages (there is no "full evaluation" as in Maple). Moreover variables are lexically scoped. This means that the following definition `f(x):=x+y;` yields a function with one argument that adds this argument to the value that y had at the time the function is compiled. There is absolutely no way to change this function, but of course, you can re-bind f to another function. You use anonymous (or unnamed) functions, like **lambda(x,x^2)**.

A second point is that the language provides closures. For instance, the function **makematrix** takes three arguments, and makes a matrix. The two first arguments indicate the size, and the last is a function that takes i and j as arguments and gives the value at positions i and j of the matrix to be computed. For instance, to make a function that adds two 3 by 3 matrices, you can write

```
mat_add(m1,m2):=
  makematrix(3,3, lambda([i,j], getelt(m1,i,j) + getelt(m2,i,j)));
```

Of course, you can add general matrices, you just need to add the code that test that both arguments are matrices, that they have same dimensions, etc. In this example, you could replace **makematrix** by another function, for instance, to print the elements of the sum without constructing it:

```
let(function=lambda([i,j], getelt(m1,i,j) + getelt(m2,i,j)),
    for a = 1 ... 3 do
        for b in 1 ... 3 do
          print (format("Element at position ~A and ~A is ~A.",
                        a ,b, function(a,b))));
```

This example shows also how local variables can be introduced, and the syntax of loops. Closures can be used in many ways. For instance, if you want to get the greatest element of the matrix, you can write the following:

```
max_of_sum(m1,m2):=
  let(function=lambda([i,j], getelt(m1,i,j) + getelt(m2,i,j)),
                                       /* The same function as before */
    let ([i=1,j=1,                     /* initial indices */
      biggest(u,v)= if u<v then v else u], /* the comparison function */
       let (next_ij()= <<if j<3 then j++   /* updates the indices */
                else if i<3 then <<i++, j:=1>>
                else error("max_of_sum","should not happen",[i,j]),
                      [i,j]>>,
            let(max=function(1,1),     /* initial value */
                repeat(8,              /* 1 -  the number of values */
                      max:=biggest(max,apply(function,next_ij()))),
                max))));
```

This example shows how to define local functions without using the **lambda** keyword, how to write conditionals, how to create begin-end structures, how to generate errors, and how to add comments to a piece of code.

A simpler way to write the loop (once **max** is initialized) is :

```
for [i,j] in cartesian_product_seq(1...n,1...m)
  do max:=biggest(max, function(i,j))
```

In this example the intervals $[1, n]$ and $[1, m]$ could be replaced by any sequence (interval, list, array, set, etc), and you can construct your own sequences, using constructors, like the cartesian product.

An important point to notice is that if a function f uses another function g, the function g has to be defined before f. This is easy to understand, but you have to take care, that if you change g, you have to redefine f, because this function is not aware of any change in the system. Of course, you can write recursive functions, and also mutually recursive ones. The last feature of the language is that you write your code into a file, and compile the file, and load it later, in this case, the order of functions in the file has no importance. Part of the library of SISYPHE (for instance, the integer factorisation package) is entirely written into this language.

4.2 Automatic Construction of Algorithms

Many mathematical structures are build with a few number of basic constructors. We have implemented in **athena** the following:

- cartesian product.
- functional exponential (applications from a model to another).
- quotient by a congruence.
- finite sequences.
- finite subsets.

These constructors very often preserve signatures and axioms, and **athena** builds automatically the algorithms of complex structures built with these constructors. We give an example (section 5) of the construction of an algebra of formal series without writing any algorithms. Many examples in various domains have been tested in practice (polynomial rings, non commutative polynomials, dioids, tensor algebras, external algebras, quotient fields, etc).

Note that for each constructor, several data structures are proposed (with one by default) e.g. dense, sparse, lazy representations for applications from a model to another. The generated algorithms are optimized and performances are comparable with built-in algorithms.

4.3 Interfaces with Other Libraries

ulysse is interfaced with several softwares, like Maple, SISYPHE (not only its library), Zicvis (a curve and surface plotter), each of them been associated to a specific model of **athena** . Other extensions are in progress.

5 Examples of Session with `ulysse`

5.1 Rewriting

We define an operator "{" associative, commutative, idempotent, and rewrite rules which express Gentzen system of sequents for propositional logic. Then we test a tautology.

```
(4) new_operator([name = "|{|":Lisp,
(4)       properties = "((is-a) (is-c) (is-id))":Lisp]);

4                                      {

(5) Gentzen:= [{#f,#F} |- {#f,#G} -> true:Boolean,
(5)      {#f and #g, #F} |- #S -> {#f,#g,#F} |- #S,
(5)      #S |- {#f and #g, #F} -> (#S |- {#f,#F})*(#S |- {#g,#F}),
(5)      {#f or #g, #F} |- #S -> ({#f,#F} |- #S)*({#g,#F} |- #S),
(5)      #S |- {#f or #g, #F} -> #S |- {#f,#g,#F},
(5)      {not #f, #F} |- #S -> {#F} |- {#f, #S},
(5)      #S |- {not #f, #F} -> {#f, #S} |- #F,
(5)      {#f => #g, #F} |- #S -> ({#F} |- {#f,#S})*({#g,#F}|-#S),
(5)      #S |- {#f => #g, #F} -> {#f,#S} |- {#g,#F},
(5)
(5)       #a <=> #b -> (#a => #b) and (#b => #a),
(5)      prove(#f) -> {H} |- {C,#f}
(5)
(5)            ]:RS;

5        ...

(6) f2:= (not (a or b)) <=> (not a) and (not b);

6                    (not a or b) <=> ((not a) and not b)

(7) rewrite(prove('f2),'Gentzen);

7                                  true : Boolean
```

5.2 Algebra of Formal Series

We build here the algebra $Q[[X]]$ of formal series in X, only with constructors of models (finite set, power, sparse and lazy representations).

```
(8) construct_model([name = {X},
(8)        constructor = oset,
(8)        enumeration = [X]]);

8                                        '{X}

(9) construct_model([name= MonX,
(9)         type = 'F_Commutative_Multiplicative_Monoid,
(9)         constructor = power,
(9)         coefficients = 'N,
(9)         generators = '{X},
(9)         operator1 = "|*|",
```

```
(9)         operator2 = "|^|",
(9)         reverse_op2 = true]);

9                                    '(MonX)

(10) construct_model([name = Q[[X]],
(10)         constructor = power,
(10)         representation = lazy,
(10)         coefficients = 'Q,
(10)         generators = 'MonX,
(10)         operator1 = "|+|",
(10)         operator2 = "|*|",
(10)         reverse_op2 = false]);

10                                   '(Q[[X]])
```

Some computations with the trigonometric series sin(X) and cos(X), mutually recursively defined with a lisp algorithm. In the lazy representation, a function beetwen models is given by a finite part of its graph, and an algorithm.

```
(11) sin:=[0,
(11)        "#.(de sinus(s ((n . x) . 1))
(11)        (selectq n
(11)         ((0 ()) 0)
(11)         (1 1)
(11)         (t (/ (lazy-coef cos (list (cons (- n 1) x))) n))))"
(11)        :Lisp]:Q[[X]];

11                              (0 + ...) : (Q[[X]])

(12)
(12) cos:=[1*X^0,
(12)        "#.(de cosinus(s ((n . x) . 1))
(12)        (selectq n
(12)         (() 1)
(12)         (1 0)
(12)         (t (- (/ (lazy-coef sin (list (cons (- n 1) x))) n)))))"
(12)        :Lisp]:Q[[X]];

12                              (1 + ...) : (Q[[X]])
```

The coefficient of X^5 in sin(X).

```
(13) 'sin((X^5):MonX);

13                                   1/120 : Q
```

Then some coefficients of cos(X) have been computed and memorized :

```
(14) 'cos;

                              2                4
14               (1 + -1/2 * X  + 1/24 * X  + ...) : (Q[[X]])

(15) 'sin;

                      1            3                5
15             (1 * X  + -1/6 * X  + 1/120 * X  + ...) : (Q[[X]])

(16) s:= 'sin*'sin + 'cos*'cos;

16                            (...) : (Q[[X]])

(17) 's((X^0):MonX);

17                                 1 : Q

(18) 's((X^10):MonX);

18                                 0 : Q

(19) 's;

19                            (1 + ...) : (Q[[X]])
```

5.3 Use of Other Softwares

```
(20) load("maple");

20                                 maple

(21) (factor(x^5+y^5)):Maple;

         2    2    4    4        3              3
23    ((x  * y  + x  + y  + (- x  * y) + (- y  * x)) * (x + y)) : Maple

(24) load("sisyphe");

24                  /u/psyche/1/ulysse/sources/sisyphe.ul
```

```
(25) integrate(1/(1+x^2+x^3),x):Sisyphe;

    =====
    \
     >                    g105 log((- 2 g105 + 3) x + (9 g105 + 1))
    /
    =====
25              3   1        1
    g105 | g105  + -- g105 - -- = 0
                   31        31

    :Sisyphe
```

6 Conclusion

We have demonstrated the applicability of our approach to a wide range of problems. The system offers a well structured mathematical environment while reasonably small and portable. Its modular construction is designed to allow easy development of further algorithms and extensions to the library. We plan now to incorporate the extensible graphic user interface developed in our SAFIR project by N. Kajler [Kaj90, Kaj91], this will make the system more accessible for other research groups.

References

[DGJ+84] J. Davenport, P. Gianni, R. Jenks, V. Miller, S. Morrison, M. Rothstein, C. Sundaresan, R. Sutor, and B. Trager. *Scratchpad.* Mathematical Sciences Department, IBM Thomas Watson Research Center, 1984.

[GGP90] André Galligo, José Grimm, and Loïc Pottier. The design of SISYPHE : a system for doing symbolic and algebraic computations. In A. Miola, editor, *LNCS 429 DISCO'90*, pages 30–39, Capri, Italy, Avril 1990. Springer-Verlag.

[INR89] INRIA. *Le_Lisp de l'INRIA Version 15.22, Le Manuel de Référence*, INRIA edition, Janvier 1989.

[Kaj90] Norbert Kajler. Building graphic user interfaces for computer algebra systems. In A. Miola, editor, *LNCS 429 DISCO'90*, pages 235–244, Capri, Italy, April 1990. Springer-Verlag.

[Kaj91] Norbert Kajler. User Interfaces for Computer Algebra: a Modular, Open, and Distributed Architecture. In *Prooceedings of the Workshop on Symbolic and Numeric Computation*, number ISBN 951 -45-5912-6 ; ISSN 0356-9225, Computing Centre, University of Helsinki, Research Reports 16, May 1991.

AlgBench: An Object-Oriented Symbolic Core System

Roman E. Maeder

Theoretical Computer Science
ETH Zürich, Switzerland

Abstract. *AlgBench* is a workbench for the design, implementation and performance measurement of algorithms for symbolic computation, including computer algebra. It provides an interpreter for a symbolic language and a skeleton for implementing data types and algorithms in all areas of symbolic computation. An object-oriented design makes incorporation of new code easy. A compiled implementation language was chosen to allow measurements of efficiency of algorithms and data structures down to machine level.
The interpreter provides a sophisticated scanner and parser for a symbolic functional language. A term rewriting system underlies the evaluator. The evaluator uses the "infinite evaluation" paradigm. Terms are transformed according to user-defined and built-in rules until no more rules match. In traditional symbolic computation systems this evaluator is a routine that dispatches on the type of the expression. In an object-oriented system this is implemented much cleaner as a virtual method. The pattern matcher and unifier is also implemented in this way. This leads to a transparent design, completely different from traditional systems. It is a strong point-in-case for object-oriented design methodology.
The data types provided are those of a symbolic system: symbols, numbers, and composite expressions (isomorphic to LISP lists and *Mathematica* expressions). Efficiency is gained by deriving subtypes of these for certain kinds of expressions (e.g. lists, pattern objects, polynomials). This system allows easy implementation and comparison of algorithms. Currently the following are investigated: Parallel/vector arbitrary precision arithmetic, parallel pattern matching, polynomials over finite fields, and interval arithmetic. *AlgBench* is written in AT&T C++ V2.0.

1 Introduction

We discuss design issues and some implementation details of our workbench for algorithm implementation. The system is not intended as yet another symbolic computation system, rather it serves the implementor of algorithms. The traditional form of publication of an algorithm in CA is a high-level description in some pseudolanguage. This form is easy to understand and sufficient for studies of correctness and asymptotic complexity. An effective implementation, however, cannot be derived automatically from such a description. It omits all questions of data representation and the practical difficulties in performing "coefficient operations" (E.g. an algorithm for factoring polynomials over finite fields might assume arithmetic in $\mathbf{F}_q$ as basic.)

AlgBench offers the following ingredients:

- An interpreter for a symbolic language. It allows easy testing of algorithms, since it can be called interactively. Test-examples can be developed easily and performance data can be collected.
- A scanner and parser implemented with lex and yacc.
- A basic set of data types and operations. These include arbitrary size integers and rational numbers, various representations of polynomials over rational numbers and integers modulo small primes.
- Functional and traditional procedural programming constructs.
- A term rewriting system, including type-constrained unification, and conditional pattern matching.
- A set of measurement tools, including timing, memory allocation statistics and tracing, inspection of internal representations, tracing of evaluation and term rewriting.

Algorithms can be written in C++. It is therefore possible to use predefined data types for the implementation of the coefficient operations or to write one's own, tailored to a particular class of hardware. Several implementations of the same specification (class definition) can be developed and compared.

The paper is organised as follows: Section 2 discusses the representation of symbolic data in comparison with the traditional LISP method of storing symbolic expressions. The implementation of type information is a first example of object-oriented design. In section 3 we briefly describe the infinite evaluator. In the next section 4 we show how inheritance allows algebraic data to be viewed in different ways, allowing both general symbolic computation and implementation of algebraic algorithms, for example for polynomials. Section 5 discusses garbage collection techniques in detail. This area is a strong point in favour of C++. A reference count scheme can be implemented in a mechanical way, refuting the claim that it is very difficult and error prone to do so. Finally, we address portability questions.

2 Data Types for Symbolic Computation

Symbolic expressions have been represented in essentially the same way ever since symbolic computation was proposed as a major application of the at that time new language LISP [McC60]. An expression is either an *atom* (a symbol, number, or character string) or is a list of expressions in the form

$$(f e_1 \ldots e_n), n \geq 0,$$

called a *normal expression*. In most systems the internal representation is fixed (linked cells in systems implemented in LISP, a contiguous array of pointers in *Mathematica*). In an object-oriented system, only an interface definition is needed to use symbolic expressions. The implementation can be changed easily. Currently,

AlgBench stores compound expressions as arrays of pointers to their elements. Symbolic algorithms need to take the storage method into account, as a comparison of the complexity of basic operations in table 1 shows.

Table 1. Complexity of operations on an expression of length n

Operation	LISP cells	array of pointers
prepend an element	1	n
append an element	n	n
get i-th element	i	1
find length	n	1
storage size	$2n$	n

The *type* or head of a normal expression is its first element. In non-LISP systems, a list $(fe_1 \ldots e_n)$ is written in functional notation as $f(e_1, \ldots, e_n)$ or as $f[e_1, \ldots, e_n]$, hence the name *head*. The head of a *symbol* is the symbol `Symbol`, the head of numbers is one of `Integer`, `Rational`, etc. In an object oriented system, `head()` is a virtual (or dynamically bound) method. This implementation method gets rid of the many dispatch tables (or `case` statements) found in traditional systems. ([RK88] used this methodology for implementing a SCHEME interpreter in C++.)

2.1 Data Representations

Here is a basic class hierarchy for symbolic expressions. This hierarchy forms the skeleton of *AlgBench*. The method `eval()` is the topic of the next section. `expr_rep` (Figure 1) is an abstract base class (it is made abstract by setting one of its methods to 0). `head()` is a virtual method. It is specialised in all subclasses. The head of

```
class expr_rep {                          // abstract base class
public:
  virtual ~expr_rep();                    // virtual destructor

  virtual expr_rep* head() const = 0;     // head of the expression
  virtual expr_rep* eval();               // evaluate: default self
};
```

Fig. 1. the abstract class expr

strings, symbols, and numbers need not be stored as an instance variable. It can simply be synthesised, for example for strings in Figure 2. where `steString` is a variable statically initialised to the symbol `String`. Strings contain an instance variable, the pointer to their character array (Figure 3):

Symbols are conveniently implemented as a subclass of strings. As an example of their properties, we show only their value (Figure 4).

```
expr_rep* string_rep::head() const
{
    return steString;
}
```

Fig. 2. specialized method for head()

```
class string_rep : public expr_rep {     // character arrays
public:
  string_rep(const char*);               // new string_rep("val")
  string_rep(const string_rep* old);     // new string_rep(old)
  ~string_rep();                         // destructor

  expr_rep* head() const;                // specialise virtual method
  const char* strval() const;            // look at its name

protected:
  char* stringdata;                      // its value
};
```

Fig. 3. the string class

For normal expressions the head is stored as element 0 of the element array. The subscripting operator

is overloaded, so that elements can be accessed as if an object of type **normal_rep** were an ordinary array (Figure 5). Features like this make C++ the preferred choice as implementation language [Str91].

3 The Symbolic Evaluator

AlgBench provides an infinite evaluator. This fits together well with its paradigm of applying rewrite rules to expressions, as opposed to applying user defined pro-

```
class symbol_rep : public string_rep {   // symbols, name is string
public:
  symbol_rep(const char*);               // new symbol_rep("name")

  expr_rep* head() const;                // specialise virtual method
  expr_rep* eval();                      // specialise virtual method

private:
  expr_rep* ownvalue;                    // value of symbol
};
```

Fig. 4. the symbol class

```
class normal_rep : public expr_rep {      // normal expressions
public:
  normal_rep(const int len);              // new normal_rep(len)
  normal_rep(const normal_rep* old);      // new normal_rep(old)
  ~normal_rep();                          // destructor

  expr_rep* head() const;                 // specialise virtual method
  expr_rep* eval();                       // specialise virtual method

  int Length() const {return length;}     // get length
  expr_rep*& operator[](int i);           // set i-th element
  expr_rep* operator[](int i) const;      // get i-th element

  normal_rep* append(const expr_rep*) const;    // append an element
  normal_rep* prepend(const expr_rep*) const;   // prepend an element

private:
  unsigned int  length;                   // number of elements
  expr_rep* *elems;                       // pointer to element array
};
```

Fig. 5. normal expressions

cedures. Rewrite rules are more flexible and are closer in spirit to how humans do mathematics. These ideas are, of course, taken from *Mathematica* [Wol91]. An infinite evaluator proceeds as follows:

- Strings and numbers evaluate to themselves.
- If a symbol has a value, it evaluates to that value, otherwise to itself.
- A normal expression $(f e_1 \ldots e_n)$ is evaluated like this:
  - The head f and the elements $e_1, \ldots, e_n$ are evaluated.
  - User defined rewrite rules are applied.
  - Built-in code is applied.
- If the expression changed as a result of one these steps, it is evaluated again.

A (one-step) evaluator in LISP is typically a dispatch routine that finds the type of the expression to evaluate and then calls an appropriate procedure to evaluate each kind of expressions (see [AS85] for an example SCHEME evaluator). Instead of a dispatch procedure our evaluator in an object-oriented system is simply implemented as a virtual method. The default implementation in the base class **expr_rep** takes care of self-evaluating expressions:

```
expr_rep* expr_rep::eval()
{
    return this;
}
```

We omit the code for the evaluation of other expressions. It is a straightforward implementation of the evaluation procedure outlined above. No code at all is necessary for strings and numbers, as these can inherit the default method.

4 Inheritance put to Work

Further development with algorithms implemented in *AlgBench* shows the importance of the concept of inheritance. The class of normal expressions is made an abstract class under which different kinds of normal expressions can be defined. One family of such subclasses are the various pattern objects used in pattern matching, not discussed further in this paper. Another subtree is the tree of polynomial representations. Consider polynomials over $\mathbf{F}_p$ for small p. By making them a subclass of `normal_rep`, these polynomials can on one hand be treated as the symbolic expressions they are, on the other hand we can still represent them efficiently, with their coefficients being stored in an array of machine `int` variables, for example.

For example, x^2+2, as a symbolic expression is thought of as `(Plus (Power x 2) 2)`. It would be stored as a coefficient array $\{1, 0, 2\}$, its degree, and a pointer to its variable. The specialised virtual method `head()` would return `Plus`, however.

Algorithms for doing arithmetic with such polynomials will be passed only the coefficient arrays as arguments, returning a new such array for the result. This array is then stored in a new symbolic expression which is then returned to the symbolic evaluator. Arithmetic with polynomials over $\mathbf{F}_p$ is so basic (among others, it is used in computing with homomorphic images, e.g. GCD and factoring of polynomials over the integers) that it should be implemented as efficiently as possible. By restricting the size of the modulus p to machine size, this can be done easily. The restriction is not serious: the product of all prime numbers $< 2^{16}$ is about $6.643 \cdot 10^{28304}$. With only a slight performance penalty, 32bit primes could be used as well.

5 Memory Management

Memory management and garbage collection techniques are an important part of the study of effective implementations of algorithms. It cannot be left to the underlying run-time system, be it Lisp or a general-purpose garbage collector. In [Dav90], J. Davenport discusses garbage collection and concludes that reference-counting is difficult to implement, requiring (preferably machine-generated) code for every variable declaration, assignment, function call, and end-of-scope. C++ provides all the tools for doing exactly this. A version of *AlgBench* with a reference-count garbage collection scheme was derived systematically as follows:

1) A first version without garbage collection was developed. All data was manipulated through pointers to objects of the symbolic class hierarchy described in section 2. Pointer variables were declared exclusively through such typedefs:

```
typedef expr_rep* expr;
typedef string_rep* string;
typedef symbol_rep* symbol;
typedef normal_rep* normal;
```

2) The definitions for these types were then replaced by a parallel class hierarchy of "smart pointers". Objects of its base class `expr` contain as only data member the (original) pointer to `expr_rep`. Overloading of the two pointer-dereferencing

operators -> and * allows them to be used exactly as if they were simple pointers. An assignment operator taking an original pointer as argument allows assignments. A constructor allows declaration of variables with initial values. A type coercion operator allows them to be used anywhere in place of a pointer, especially in these constructors and assignment operators (Figure 6). The sub-

```
class expr {
public:
    expr(expr_rep *obj);                    // expr e = new expr_rep();
    void operator=(expr_rep *obj);          // e = e1;
    expr_rep* operator->() const { return e_ptr; }  // e->
    expr_rep& operator*() const { return *e_ptr; }  // *e
    operator expr_rep*() const { return e_ptr; }    // type conv
protected:
    expr_rep* e_ptr;                        // the pointer itself
};
```

Fig. 6. smart base class pointers

classes simply cast the pointer to the corresponding subclass of **expr_rep**. The assignment operator merely calls the operator of its base class (assignment operators are not inherited), and likewise for the constructor (Figure 7). None of the

```
class normal : public expr {
public:
    normal(normal_rep *obj) : expr(obj) {}
    void operator=(normal_rep *obj) {expr::operator=(obj);}
    normal_rep* operator->() const { return (normal_rep*) e_ptr; }
    normal_rep& operator*() const { return *(normal_rep*) e_ptr; }
    operator normal_rep*() const { return (normal_rep*) e_ptr; }
};
```

Fig. 7. smart subclass pointers

methods needs to be virtual, causing no storage overhead. An object of a smart pointer class occupies only as much space as the pointer itself. Most methods can be inlined.

3) The class **expr_rep** is augmented by a reference count instance variable and (inline) methods for incrementing and decrementing it. If the reference count reaches 0 after decrementing it, the object is freed, using the virtual destructor **~expr()** (Figure 8).
4) A default and copy constructor is added to the smart pointer classes, as well as a destructor. These, and the assignment operator, are coded to perform the changes in the reference counts of the objects pointed to. As an example, Figure refF:assign shows the assignment operator: The order of the first two state-

```
class expr_rep {
  expr_rep() : refcount(0) {}    // constructor
  virtual ~expr_rep();           // virtual destructor
  ...
  void RefIncr() {refcount++;}
  void RefDecr() {if (--refcount == 0) delete this;}
protected:
  int refcount;
};
```

Fig. 8. Reference count methods

```
void expr::operator=(expr_rep *obj)
{
    obj->RefIncr();
    e_ptr->RefDecr();
    e_ptr = obj;
}
```

Fig. 9. assignment operator

ments is important. Decrementing first is wrong when assigning to the same variable (`e = e`) if the reference count happens to be 1. The object would be freed. Testing for null pointers can be avoided by initialising all pointers to the address of a special object instead of null.

The C++ compiler treats function arguments and return values correctly by using the appropriate copy constructors where necessary. Since no uncounted references ever exist, a correct compiler guarantees the correctness of this memory management scheme. It is worth pointing out that no change to the code of the system itself was necessary.

6 Portability

The standardisation of C (ANSI Standard, [KR88]) and the standard libraries (especially the streams library) provided with AT&T C++ V2.0 makes C++ programs very portable. This is a dramatic improvement from the situation with the "old" C described in [Dav90]. *AlgBench* has been compiled and tested on Sun-3, SPARC (both under SunOS 4.1), Convex C220 under Convex OS V9.0, Alliant FX/80 under Concentrix 5.6, a Sequent S81b under Dynix 2.0, and a Cray Y-MP under Unicos 6.1. The only machine-dependent features are the word size and integer data types used for implementing arbitrary size arithmetic.

Acknowledgements

Contributions to the design and implementation of *AlgBench* came from G. Grivas, S. Missura, J. Rudnik, and F. Santas. Students in my course "Design and Implementation of Symbolic Computation Systems" at ETH Zürich wrote their own symbolic systems, from which occasional ideas have been taken.

References

[AS85] H. Abelson and G. J. Sussman. *Structure and Interpretation of Computer Programs.* The MIT Press, Cambridge, Mass., 1985.

[Dav90] J. H. Davenport. Current problems in computer algebra systems design. In A. Miola, editor, *Design and Implementation of Symbolic Computation Systems (Proceedings of DISCO '90)*, volume 429 of *SLNCS.* Springer Verlag, 1990.

[Han90] Tonly L. Hansen. *The C++ Answer Book.* Addison Wesley, 1990.

[KR88] B. W. Kernighan and D. M. Ritchie. *The C Programming Language.* Prentice-Hall Software Series, Englewood Cliffs, New Jersey, second edition, 1988.

[McC60] John McCarthy. Recursive functions of symbolic expressions and their computation by machine I. *Journal of the ACM*, 3:184–195, 1960.

[RK88] V. F. Russo and S. M. Kaplan. A C++ interpreter for scheme. In *Proceedings of the C++ workshop.* Usenix Assoc., 1988.

[Str91] Bjarne Stroustrup. *The C++ Programming Language.* Addison Wesley, second edition, 1991.

[Wol91] Stephen Wolfram. *Mathematica: A System for Doing Mathematics by Computer.* Addison-Wesley, second edition, 1991.

SYMO2: Objects and Classes for Symbolic Computation Systems

Philip S. Santas

Department of Computer Science
ETH Zurich, Switzerland
email: santas@inf.ethz.ch

Abstract. SYMO2 is an object-oriented environment built in various symbolic computation systems for the examination of their behaviour under the combination of different knowledge representation techniques. SYMO2 is not an interpreter built in the examined symbolic languages but a powerful collection of functions, which adds advanced programming features to them and enlarges their representation capabilities.
With the usage of SYMO2 direct comparison of symbolic computation systems is possible, since it embodies to each of them characteristics from the other systems in which it is implemented. By allowing the introduction of object-oriented characteristics like classes of objects, metaclasses, multiple inheritance, message passing, behaviour overriding and enrichment, data encapsulation, concurrent methods, and run-time behaviour modifications to these systems, schemas, which have received applause in the field of symbolic and algebraic computation, can be directly combined and evaluated. Such schemas include data types, domains, properties, intervals and parallel execution of algorithms. Furthermore new characteristics like multi-level metaclassing, declarations and uniform manipulation of objects are introduced.
Results from the comparisons performed by SYMO2 can be useful for the design and implementation of new, more effective and consistent and better structured symbolic and algebraic computation systems.
SYMO2 is currently implemented in Maple, Mathematica and Scheme.

1 Introduction

Object-oriented systems are believed to be of considerable value in domains such as software engineering [Cox86] computer graphics [Goldberg 89], office automation [Nierstrasz 89, Banerjee 87] and artificial intelligence [Coiffet 90]. They combine with clarity well known techniques (modularization, data abstraction) with new concepts, and give a new framework for the implementation of typed and distributed applications, code reuse, rapid prototyping and teaching of programming concepts.

Object-oriented languages usually exhibit the following characteristics: (a) The traditional subdivision of a system into data (Packages, Modules) and programs (processes) is replaced by the notion of objects, entities that encapsulate the data and the allowable operations (methods, messages) on them and (b) Objects are grouped and structured into classes which in turn can be subclasses or classes of classes (metaclasses), from which they inherit their properties and behaviour.

There are advantages to enhancing this classical definition of objects and inheritance with the properties of consistency, concurrency and autonomy: classes define behaviour propagated to their subclasses and all their instances but only them can inherit it. On the other hand many objects can be active at the same time being manipulated through a possibly concurrent message passing mechanism.

Because of its characteristics, object-oriented programming can be useful in functional and symbolic computation [Fateman 90]; this can be shown by the incorporation of such ideas into major symbolic systems:

CLOS [Moon 89] is an object-oriented programming language which is embedded in COMMON LISP [Steele 84]. CLOS is characterized by multiple inheritance, generic functions, and a separation between structure and behaviour. In [Sussman 85] object-oriented programming techniques with SCHEME are presented; they do not include classes and inheritance and the desired polymorphism is achieved by message passing performed by dispatch of operations. Symbolic Computation Systems built in LISP like MACSYMA [Mathlab 85] and AXIOM [Jenks 85] can be influenced by such advances in Lisp-like environments.

[Wolfram 88] assumes object-oriented characteristics in MATHEMATICA. This claim is based on the possibility of polymorphic typed-based dispatch of operations performed in type matching, but MATHEMATICA supports neither classes nor inheritance [Fateman 91]. MAPLE [Geddes 90] on the other hand seems not to follow the systems above and as a programming environment it does not provide even the facilities of polymorphism in the sense that it does not allow multiple incremental definition of functions with the same name operating on different number or type of arguments.

SYMO2 is an experimental programming environment for adding the desired object-oriented characteristics to the symbolic systems presented above. SYMO2 borrows concepts from Smalltalk [Goldberg 89], CLOS, and Miranda [Turner 85] and is implemented for our purposes in MAPLE, MATHEMATICA and SCHEME.

SYMO2 is not a general purpose object-oriented language, but rather an application environment for symbolic computation systems. Its main purpose is the studying of the behaviour of the symbolic systems when concepts like data types, inheritance and operations like declarations are introduced. Comparisons between the traditional and the new type-, object- or domain-based symbolic and algebraic computation systems have been already presented [Fateman 90], but these results cannot be generalised since different systems based on different philosophies and with different targets are compared. SYMO2 is built within the systems of interest and their behaviour can be directly examined under the new introduced concepts.

2 What is Object Orientation in Symbolic Computation?

Before starting the presentation of SYMO2's structure, we examine an important point concerning the new generation symbolic computation system Axiom and some research dedicated to the introduction of object oriented features to algebraic computation [Fortenbacher 90, Limognelli 90] and explain the motives which pushed us to follow a rather different approach to class based semantics.

2.1 Subclasses vs Instances

AXIOM has been characterised as object oriented and this point has been documented with the support of classes, multiple inheritance and abstract parameterised datatypes [Jenks 85, Davenport 90]. The terms Categories and Domains are borrowed from Modern Algebra in order to model semantics, which in usual Object Oriented languages would have been represented as instances of Classes (in [Davenport 90] Categories describe Algebras); the instances of metaclasses form classes by themselves: in other words Categories and Domains are used for the modelling of Metaclasses and Classes.

We adopt the modelling of [Davenport 90] for its correctness in representing algebraic concepts such as Rings and Fields, in contrast to [Limognelli 90] which fails to distinguish between subclasses and instances:

The subclass relation (denoted with $\prec$ in the rest of the text) is an irreflexive and transitive binary ordering relation in the partially ordered set of classes. For the classes x, y the expression $x \prec y$ stands for x is subclass of y or an natural language *every x is a y*. The instance relation (denoted with $\in$ for emphasis to the semantics) represents the membership in a set and as such it is not transitive (irreflexiveness is obvious): If $(a \in b) \wedge (b \in c) \not\Rightarrow a \in c$. The negative subclass relation (denoted with $\not\prec$) is the negation of the subclass operation. For the classes x,y the expression $x \not\prec y$ stands for x is not subclass of y. Obviously $a \in b \Rightarrow a \not\prec b$.

Using this notation we can model simple algebraic concepts: as an argument we elaborate the examples from [Limognelli 90]:

$$AbelianGroup(+) \prec Group(+) \prec Monoid(+)$$

$$(Ring(+,*) \prec AbelianGroup(+)) \wedge (Ring(+,*) \prec Monoid(*))$$

The use of Multiple Inheritance in this example over Single Inheritance is an advantage for the modelling of Algebraic Concepts as it is shown in [Santas 92]: Class $Ring(+,*)$ inherits from $AbelianGroup(+)$ the operation $+$ and from $Monoid(*)$ the operation $*$. The usage of non-strict inheritance as this has been defined in [Limognelli 90] is out of the purposes of this example.

The members of the class Ring form its instances, which can be the classes of Polynomials, Matrices, Integer etc:

$$Integer \in Ring$$

$$Matrix \in Ring$$

$$Polynomial \in Ring$$

The class Integer is definitely not a subclass of the class Ring since conceptually the two classes do not share any common structure or operation: Integers can be added to each other, like Matrices or Polynomials do, but Rings cannot: the operations of addition and multiplication defined in the class Ring are to be used by its instances; furthermore they can be inherited by its subclasses like the class Field:

$$Field(+,*) \prec Ring(+,*)$$

To be even more precise, Polynom is not even class, but can be modelled as a function, which returns, say, Polynoms over the Integers: $P(\mathcal{Z})$. In this sense as it has been shown in [Fortenbacher 90], Integers form a subclass of $P(\mathcal{Z})$, since there is a coercion from $\mathcal{Z}$ to $P(\mathcal{Z})$:

$$Integer \prec P(Integer) \text{ or more formally } \mathcal{Z} \prec \mathcal{P}(\mathcal{Z})$$

The distinction between subclasses and instances is important for the correct modelling of algebraic concepts. We come back to this subject in section 5.

2.2 Multi-Level Metaclassing and Instantiation

The model of Single-Level Metaclassing (Categories and Domains) is not very strong for modelling certain Symbolic Computation problems. We elaborate a case from Interval Manipulation, which is a handicap for Symbolic Computation and Artificial Intelligence:

Interval Arithmetic has been employed for the solution of problems involving Qualitative Information in Qualitative Physics or Properties and Inequalities in Algebraic Computation and Symbolic Integration.

Real Intervals are presented as Ordered Pairs $< a, b >$ with $a, b \in \mathcal{R}$ and $a \leq b$. A Closed Real Interval $< a, b >$ includes all the real numbers between a and b (together with a,b). Obviously Intervals form Subsets of the Set of Real Numbers, and they can be represented as subclasses of Real, although not in the same sense as Integer is a subclass of Real: intervals are not Subrings of Real while Integers are. In this way an instance of the class Interval is automatically a subclass of the class Real (real numbers can be regarded as the open interval $< -\infty, \infty >$

The Interval Addition and Subtraction are defined as following:

$$< a, b > + < c, d > = < a + c, b + d > \text{ (ex: } < 0, 1 > + < 0, 1 > = < 0, 2 >)$$
$$< a, b > - < c, d > = < a - d, b - c > \text{ (ex: } < 0, 1 > - < 0, 1 > = < -1, 1 >)$$

It can be shown that Intervals with the addition operation form a Monoid where $< 0, 0 >$ is the 0 element. However they cannot form a Group due to the lack of inverse for Intervals of the form $< a, b >$ with $a \neq b$. Relevant occur with multiplication, where $< 1, 1 >$ is the Unit, and there is no inverse for intervals other than $< a, a >$ with $a \neq 0$.

More formally:

$$RealInterval \in Monoid(+)$$

$$RealInterval \in Monoid(*)$$

$$< a, b > \prec Real, if a, b \in R$$

$$< a, b > \in Interval$$

The class RealInterval is by its own right a Metaclass, the instances of which are subclasses of the class Real.

To make things even more dramatic, consider the Interval $< -1, 1 >$, the elements of which form a Monoid in respect to multiplication (this interval is very useful for the modelling of the sinus and cosinus functions).

$$< -1, 1 > \in Interval$$

$$< -1, 1 > \in Monoid(*)$$

Obviously the multi-level Metaclass representation in these examples cannot be reduced into single metaclassing without loss of the information passed through the relationships among classes and their instances. Furthermore such reduction may lead to strange mathematical models [Santas 92].

2.3 Behaviour Representation vs Implementation

In the object oriented paradigm in general and Axiom in particular it is assumed that specific operations are not allowed unless previously defined in a certain context [Fateman 90]. In this rational every class should define operations permitted only for its instances. One would expect from AXIOM that any variable, which is not assigned to any object and is not a number, or element of a ring in general, or not declared as such, would not be accepted as argument of the operation +. In other words, if the system holds addition or any arithmetic operation for elements of a ring, then all these objects and only these should be able to participate as arguments to this operations. Relevant rational can be assumed for elements of ordered sets and operators for comparison.

Axiom treats all undeclared variables as Indeterminates. As such they accept addition and multiplication, while common simplification is possible. It is out of the scope of this paper to discuss the implications of this otherwise well motivated treatment. On the other hand assumptions, which have nothing to do with polynomial operations, are made about their magnitude while comparing them with 0 or any other number: the lexicographical ordering among the Indeterminates causes them to be Positive and bigger than any number! Even worse, lexicographical ordering is applied to Complex numbers, although Complex numbers are not an ordered Field.

These case examples show that the argumentation in [Fateman 90] can be overridden by certain design decisions, if there is no clear distinction between modelling of behaviour and implementation: hacker approaches can be the result of addition of inheritance links which have nothing to do with knowledge representation (for the reader's information we note that the above behaviour is not occurring in systems like MATHEMATICA or MAPLE although their design is not based on classes).

2.4 Symbols and Values: Some Integers are more Integer than others

Starting a session with the representative of the class based algebraic systems, we observe that when a variable is declared as member of an instance of an ordered ring, errors are prompted during the evaluation of operations fully supported by the behavioural representation:

```
->a:Integer                                              Type: Void
->a+a
   a is declared as being in Integer but has not been given a value.
->b:PositiveInteger                                      Type: Void
```

```
->b>0
   b is declared as being in PositiveInteger but has not been given a
   value.
```

Obviously Axiom does not permit pure symbolic manipulation of entities and expects a value from its objects:

```
->a:=3
   (5)  3                                       Type: PositiveInteger
->a+a
   (6)  6                                       Type: PositiveInteger
->a>0
   (7)  true                                          Type: Boolean
```

The above examples show that AXIOM handles objects of the class Integer in two different ways, discriminating against those which have no value, by not allowing them as valid arguments of normal arithmetic operations and comparisons, while it allows undeclared variables to do the same thing.

This handling contradicts the classical mathematical concept of the class, the members of which are expected to follow uniform behaviour in certain relations [Peano 97, Russell 10].

Furthermore the type declaration has been degraded to a command for discouraging the programmer from doing certain assignments (the same complaints can be directed to other polymorphically-typed functional languages like ML [Milner 90]), when it should actually be a powerful operation for the defining of the membership of a variable in a class, and consequently to allow a real symbolic manipulation of concrete concepts of mathematics. Obviously this treatment has to do with implementation, and is another sign of the trade-offs while sacrificing correctness of semantics to implementation.

2.5 Putting Classes, Metaclasses, Instances and Declarations together

In SYMO2, declarations create new objects and introduce the variables into the members of a class; refined behaviour of the variables occurs based on the context defined by each class. Metaclasses stand for the creation of templates for algebraic operations and their properties, while their implementation is to be found in the class' instances. The whole package cannot be compared in complexity with already existing systems, but directions for consistent object oriented design and separation of the knowledge representation from the implementation can be outlined.

In the following sections the basic characteristics of SYMO2 are discussed with special notation to the semantics and the implementation useful for the symbolic systems under study. Simplified source code in MAPLE V, MATHEMATICA 2.0 and SCHEME 7.1.0 (BETA) is provided according to the language support for specifications and the usefulness with which the chosen characteristics are implemented in each system.

3 Overview of the SYMO2 System

This section briefly introduces the features of the SYMO2 language relevant to the issue of schema evolution.

3.1 Types and Classes - Inheritance

SYMO2 uses semantics from class based object-oriented languages. A class defines the structure and the behaviour of a set of objects called its instances. Not all of the objects belong to a class, making SYMO2 a hybrid system. Objects are the values of all the variables in SYMO2.

- The structure of an object is defined
  - by a type. This holds particularly for built-in objects like Integers, Reals, Strings, Symbols, Lists, Tables, etc. Such objects may include any data defined in the already existing Packages of MATHEMATICA or MAPLE; their typical behaviour is not very much influenced by the introduced extension.
  - by reference to a class, a set of variables which reflect its internal state, and a set of operations which manipulate these variables.

Classes are related to each other through inheritance links. A class inherits dynamically the structure and behaviour of its superclasses. SYMO2 supports multiple inheritance, like [Moon 89, Jenks 85] in contrast to [Abdali 86]. Multiple inheritance is the best choice for a symbolic computation system, since it expresses properly the relationships among the algebraic types, which in general form a graph and not a tree. It could be quite unorthodox and troublesome for designers and mathematicians to emulate the relationships among the various types of Sets, and the instances of Rings by using single inheritance [Santas 92].

Multiple inheritance is introduced in SCHEME by the following simplified code:

```
;; multiple inheritance

(define empty-environment ())

(define NewClass
  ;; A Class holds its direct superclasses and
  ;; a list of methods defined in it (the environment)
  (lambda superclasses
    (cons superclasses empty-environment)))

;; access functions for classes
(define SuperClasses car)
(define Methods cdr)

(define (addMethod class name function)
  ;; Addition of a new method within a class at run time
  (set-cdr! class (cons (cons name function) (Methods class))))
```

```
(define (addSuperclass class superclass)
  ;; Addition of a superclass in run time
  (set-car! class (cons superclass (Superclasses class))))

(define (NewObject Value Class)
  ;; Constructor for Objects.
  ;; Value is a list of the values of the instance variables of the object
  ;; Class is the class in which this object belongs.
  ;; The description of the concurrency mechanism is omitted
  (define (lookup-method name cl)
    ;; search method in class cl and its superclasses
    ;; returns method's function if found; else: ()
    (define (lookup-in-classes list-of-classes)
      ;; lookup in each class of the list
      (if (null? list-of-classes)
          ()

          (let ((fun (lookup-method name (car list-of-classes))))
            (if fun fun
                (lookup-in-classes (cdr list-of-classes))))))
    (if (eq? name 'value)
        (lambda (x) value)
        (if cl
            (let ((method (assq name (Methods cl))))
                (if method (cdr method)
                    (lookup-in-classes (SuperClasses cl)))))))
  ;; Message Passing mechanism
  (lambda (name . args)
    (let ((fun (lookup-method name class)))
      (if fun (apply fun (cons value args))
          (write "message not understood -- lookup")))))
```

A class definition introduces (a) a set of instance variables, which together with the instance variables inherited by its superclasses define the structural template-like implementation of its objects, and (b) a set of methods which together with the inherited methods from its superclasses define their behaviour.

3.2 Instance Variables

Instance variables are stored by name inside a class, while their values are stored in the objects of that class; they are manipulated via a small set of methods enabling a class to create and destroy dynamically instance variables. The unique storing of the name of the instance variables within a class saves a lot of space, since these symbols are not duplicated in every object, as it happens in some Scheme implementations; their values are accessed via indexes. This approach has no negative effects on the execution speed of programs. Their equivalent in MATHEMATICA semantics are the elements of a normal expression.

Generally, most standard variables are created and set default values at the birth of an object and are important for the construction of coercion functions: in a schema allowing Integers to inherit from Polynomial[Integer], the length of the list of coefficients should be 1 for objects of the class Integer when they coerce.

In MATHEMATICA the dynamic addition of instance variables has the simplified form:

```
AddInstanceVar[class_,instVars_List] :=
        (* Addition of a List of instance variables *)
 Block[
  {clInVar=AllInstanceVars[class],
        (* includes all the instance variables of the superclasses *)
   inVar, (* includes the instance variables of instVars which are
            not already included in class or in its superclasses *)
   allSub=AllSubclasses[class],
   i},
   inVar=Complement[instVars,clInVar];
   If[inVar!={},
        JoinTo[class[InstanceVars],inVar];
        AllInstanceVars[class];
        (* Update the list AllInstanceVariables of the class class *)
        Do[ UpdateInstanceVars[allSub[[i]]];
           (* Removal from the definition of the subclasses the
             instance variable which is introduced in the class *)
            AllInstanceVars[allSub[[i]]],
              {i,1,Length[allSub]}],
    ];
   class]; (* Return the class *)

AddInstanceVar[class_,instVar_Symbol] := class /;
          MemberQ[AllInstanceVars[class],instVar];
  (* Add a single instance variable *)

AddInstanceVar[class_,instVar_Symbol]:=
          AddInstanceVar[class,{instVar}];
  (* Add a single instance variable.
     Massive addition of instance variables should be encouraged *)
```

In reality various techniques which speedup the updating of the instance variables within the objects are used: some of the work to be performed during the addition or the removal of instance variables can be delayed, or is not performed at all for objects which do not use the new instance variable. More details are discussed in section 4.

3.3 Methods

Methods form the executable part of an object. Their definition is similar to that of a procedure: they receive parameters, perform some computation and, return a meaningful value to their calling environment. They can have side effects and modify the state of their arguments.

Methods serve additional purposes that distinguish them from ordinary procedures:

- They can be used as code modules like the rules defined in MATHEMATICA with the virtue that they are better controlled, they are sorted automatically according to the inheritance graph and when supported directly by the interpreter they can be more efficient (in this sense they are evaluated faster in MAPLE and SCHEME than in MATHEMATICA.
- They are responsible for the formation of templates for classes when they are found in Metaclasses. In this sense the operator Plus defined in class Ring forms a schema for the implementation of addition for Integers by providing the properties of commutativity (Orderless), associativity (Flat), etc.
- They are important as targets for the message passing facility
- They determine the lowest level of concurrency in the system: an object executing a method is guaranteed not to be interrupted by an incoming message.

There are two categories of methods: static and dynamic.

A static method is defined by a class and is inherited by all its subclasses. In the case that inherited classes define methods with identical names, the SYMO2 system builds an overriding mechanism for MAPLE and a refined set of rules for MATHEMATICA, which offers several advantages as we will see in the next section. Generally, this implies that some base method can be executed when the relevant method in the subclass has failed (like the rule application mechanism in MATHEMATICA.

The implementation of a static method is not available to the instances of a class; therefore, static methods cannot be transferred to foreign contexts by moving objects: this makes method compilation [Deutsch 84] relevant to the one applied in Smalltalk or Axiom attractive, and this option should be of consideration for system developers. Furthermore, the exact implementation of a static method may vary in different contexts, even if its interface (name and formal parameters) remains the same.

Whereas static methods pertain to the class that defined them, dynamic methods belong to the individual objects - which are entitled to add, modify or drop them freely. Objects carry dynamic methods with them when they migrate to other contexts. Such facility is important for the modelling of various finite data structures like finite Rings which share common 0- and 1-elements. Obviously built-in objects do not define dynamic methods.

Since objects can form classes by themselves (Integer is an Instance of Ring), dynamic methods obey to the class mechanism: they can be inherited, and they can be exchanged between objects. A dynamic method can be complement and even override a static one, without any risk for abnormal behaviour of objects or class-objects.

The instantiation mechanism can be considered as a form of dynamic method creation: Although class Integer is a subclass of the class Real, it is an instance of the class Ring (Real is instance of the class Field): the definition of multiplication in the class Ring and its connection to addition, is transferred through the instantiation relationship to the class Integer and it overrides the enriched definition of multiplication inherited from the class Real. Stronger assumptions can be done for the Real Intervals which form subclasses of the class Real in which subtraction is defined in a quite different way.

As far as the object (or the programmer) is concerned, there is no difference between invoking a dynamic method, a static one or a combination of both; the invocation mechanism handles all these situations transparently. For class-objects, the structure of which is not expected to change and is well documented by algebra theory, the possibility of method compilation and linking remains attractive.

The ability to mix static and dynamic methods provides a good starting point for developing interesting SYMO2 applications and experimenting with structures like the ones proposed above or in [Abdali 86, Ungar 87, Fateman 90].

3.4 Polymorphism in MAPLE, Rule Ordering in MATHEMATICA

Inheritance in SYMO2 is based on behaviour overriding (or refinement) and enrichment. In MAPLE overriding is coded as:

```
addMethod :=
    proc(class,method,code)
        method(class) := code;
        appendTo('class(methods)',method);
        class
    end;

appendTo := proc(a,b) (a := [op(eval(a)),b]) end;

New := proc(obj,class)
        Class(obj) := class;
        assign(obj,subs(OBJ=obj,
          proc() op(0,args[1])(findMethod(Class(OBJ), op(0,args[1])))
                   (procname,op(args[1])) end))
       end;

findMethod := proc(cla,meth)
                if member(cla,Classes) then
                    if member(meth,cla(methods)) then cla
                    else findMethodInList(cla(superclasses),meth)
                    fi
                else NILCLASS
                fi
            end;
```

This simple mechanism adds incremental polymorphic characteristics to MAPLE and SCHEME by allowing different definitions of a method within different classes. In SYMO2 this feature is more complicated and multiple definitions of a method with the same name are allowed within one class: this resembles to multiple function definitions in MATHEMATICA.

Enrichment is an advanced characteristic of object-oriented systems like SMALLTALK implemented by using the variable 'super'. In SYMO2 the concept of enrichment is taken from MATHEMATICA's rule evaluation sequencing: a method defined in a subclass does not necessarily override the method defined in the superclass but adds behaviour (code) to it.

This feature is very useful in the case of simplifications: class Complex may define simplifications valid for all its elements, but class Real can enrich them even further as it has been shown in [Bradford 92].

SYMO2's enrichment mechanism is presented in MATHEMATICA as following (the provided simplified code assumes single inheritance):

```
Attributes[EnrichMethod]={HoldAll};

EnrichMethod[class_[method_[args___]],code_]:=(
            (* Add the method method with arguments args
               in class class *)
 (class[method[self_,args]]:= code);
            (* add the enrichment mechanism for this method
               at the end of the class definitions *)
  AppendTo[DownValues[class],
       Literal[class[method_[self_,ANY__]]] :>
         ListDo[supClass[method[self,ANY]],
               {supClass,class[Superclasses]}]
      (* ListDo[expr,{localVar,list}] evaluates expr
         Length[list] times where localVar is
         the (i++)th element of list each time *)
  ] )
```

Operation enrichment occurs when a class defines a method with the same name as one provided by any of its superclasses. In this case, the method defined in the subclass can have a different number of arguments and execution preconditions than the one of its superclass. Moreover the types or the classes of the method's arguments specified in the subclass can be different from those specified in the superclass. This new mechanism can be seen as a partial solution to MATHEMATICA's rule ordering problem [Maeder 91].

3.5 Privacy of Data - Concurrency of Message Passing

Instance variables can be private to an object by not being visible from other objects, but it should be permissive to send explicitly their description or their copies to another object. In this sense, each method-call creates a local environment, where operations on instance variables of the object receiving this method are allowed only from that particular object.

Concurrency of message passing [Tomlinson 89, Neusius 91] involves the definition of the global variable ActiveObjects and the instance variable MessageQueue which is included in all classes. Functions and methods for queue-manipulation are provided. When an object receives a message, it stores it at the end of its private

MessageQueue and it executes the relevant method when all the other messages send to this object are evaluated. Moreover, the ability to decide the circumstances under which a communication can take place with their acquaintances reinforces the autonomy of objects and their ability for evaluating methods concurrently. This approach has a positive impact on the evaluation of symbolic computation algorithms where operations like pattern matching can be parallelised.

Both static and dynamic methods are public and can be sent from any object in any environment. There seems to be no reason for symbolic algebraic applications to add private methods, an addition which increases the complexity of the entire system, while the security gains are minimal, since algebraic algorithms tend to be by nature side effect free.

Private instance variables and concurrent methods are non standard features of SYMO2: they increase the system's complexity and running time of programs, and there is no supporting environment in MAPLE and MATHEMATICA where concurrency issues can be evaluated. Functional languages in general oppose encapsulation of data [Moon 89]. On the other hand SCHEME allows concurrent evaluation, therefore this characteristic is included only in the SCHEME implementation of SYMO2. A more primitive form of message-queues which allows the studying of message priorities is implemented in MATHEMATICA, where the attribute Orderless can emulate partially a non-deterministic environment.

4 Schema Modification

In this section we present the schema modifications supported by the SYMO2 environment and we outline how they are processed. These modifications reflect the designer's intention to start with simple but fundamental and useful modifications in order to understand their impacts on the schema, its implementation and its associated knowledge base. Modifications can be temporal or permanent. Temporal modifications are performed very fast, do not have global effects, and allow the system to return to its previous state. Permanent modifications are performed with a user command and they change totally the structure of classes and their instances.

4.1 Modifications of the Class Contents

This category contains the addition, deletion and renaming of instance variables and methods, and it is mostly related with the implementation of concepts rather than their modelling. The possibility of having a tremendous amount of classes in the system for the modelling of concepts while there are few data structures which implement them [Santas 92], dictates the careful storage of instance variables.

The class should not redefine an already inherited instance variable with the same name. If the class previously inherited a feature with the same name, the induced redefinition is ignored.

Temporal addition of an instance variable has no immediate effect on the structure of the objects of that class. An object, which assigns a value to the new instance variable, adds this value in a certain position to the list where this object keeps the values of its instance variables. The rest of the objects do not need to add an empty slot to their corresponding list, if they do not need this instance variable. In this way a class is able to add with the minimum cost of time new temporal behaviour (like exception handling) to its instances. This characteristic is of major importance in computer algebra, since one has to find an effective and efficient representation of various data structures like Integers and Polynoms when they are interact together.

Addition of a method in a class poses no particular difficulties, and specialises or enriches the behaviour of a class. Methods can be copied from one class to another if this is desired, but there is a more powerful way of emulating the results of such action as we will see in section 4.2.

Removal of an instance variable can be performed only in the class defining it. The deletion is accepted and propagated to every subclass that inherits the instance variable without inheriting it from another superclass. If the instance variable to be deleted is inherited from a superclass, or is not included at all in the class definition, the modification is ignored. Temporal removal of an instance variable has no effect on the structure of the objects.

4.2 Modification of the Inheritance Graph

This refers to adding and removing a class as well as adding and removing an inheritance link between a class and a direct superclass. Inheritance graph modifications are fundamental and they cope with the general architecture of the application (i.e. the concepts introduced and their relationships). Temporal inheritance graph modifications are the most commonly used modifications in SYMO2 since one needs to transfer classes of objects (like polynomials or integers) from one domain to another [Fateman 90].

- **Create a Class**

Both temporal and permanent creation of a class as a leaf of the inheritance graph are implemented in almost the same way, since no class instances are involved which could increase the execution time. Adding a class in the middle of the inheritance graph can be achieved by a combination of a class creation and superclass addition. The name of the class must not be used by an already defined class, otherwise the creation is ignored. The superclass(es) specified must have been previously defined.

The innovative facility offered by SYMO2 in modification issues concerns the flexibility in the ordering of class creations. Although the environment requires a class to be created before its subclasses, it does not constrain the classes appearing in the specification of a feature of the new class to be already defined. This allows the programmer to develop and test a design and its domains step by step, leaving slices of the inheritance graph undefined while testing others.

- **Delete a Class**

This modification can be applied only to the leaves of the inheritance graph. Class deletion in the middle of inheritance graph can be achieved by a combination of inheritance link deletion and class deletion. All the instances of this class are removed from the system. The implementation of class deletion in MATHEMATICA follows:

```
RemoveClass[class_]:=(
    AddSuperclassesTo[class[Superclasses],class[Subclasses]];
    AddSubclassesTo[class[Subclasses],class[Superclasses]]
    RemoveSubclassesFrom[{class},class[Superclasses]]
    Map[RemoveObject,class[Objects]];
    RemoveFrom[Classes,class];
    Clear[class]
        (* Remove all the methods associated with class *)
    ) /; MemberQ[Classes,class];
```

– **Add an Inheritance Link to a Class**

Addition of an inheritance link is useful not only for the definition of classes in the middle of the class inheritance; more important it serves for the algorithmic manipulation of various data structures in run time. Consider the class Complex, which is coded as instance of the class Field. If one wants to use methods from the vector analysis in order to solve problems involving complex numbers, a very effective solution is the addition of the class Vector[Real,Real] (2-dimensional vectors) in the superclasses of the class Complex. Obviously this addition should not introduce new instance variables in the implementation of the class Complex.

Permanent addition of inheritance links is performed in two steps: SYMO2 checks if the new inheritance link induces any cycle in the inheritance graph. Then the features provided by the new superclass (either inherited or locally defined) are propagated along the new link and the same operations as for instance variable addition are performed.

In a temporal addition the above operations are avoided and only the instance variables of the new superclass are added to the class. The new superclass has priority over the previous ones in the execution of methods which are not defined in the class. This order is based on the rational that a class of objects need to change context, because these objects are expected to behave in a different way than they were before: the new behaviour must have priority over the standard one. Obviously name conflicts are resolved by these priorities. Adding temporarily a whole superclass support is more effective technique than the temporal addition of instance variables or methods in a class.

– **Remove an Inheritance Link from a Class**

Temporal removal of an inheritance link does not affect the instances of a class and is allowed only after a previous temporal addition of this link.

Removing permanently a superclass from a class can lead to schema inconsistencies: some method definitions in the class or in any of its descendants might have used instance variables which were previously inherited through this link. Therefore all these instance variables are reintroduced in the class. The same process is not assumed for methods which are inherited through this link.

5 Declarations, Operations and Properties

As noticed in a previous section, SYMO2 makes no distinction between objects of the same class: their overall behaviour is uniform. For the class declaration the Declare function has to be applied (in MATHEMATICA 2.0 it is possible to change the syntax and use the operator :: instead). A declaration creates an object with a behaviour determined from the class in which it belongs:

```
In[2]:=  a :: Complex
Out[2]= (a)::Complex
In[3]:= a+a
Out[3]= (2 a)::Complex
In[4]:= a-a
Out[4]= (0)::Complex
In[5]:= (b::PositiveInteger) > 0
Out[5]= True
In[6]:= a=3
Out[6]= (3)::Integer
In[7]:= a+a
Out[7]= (6)::Integer
```

In the 4th output we notice the notion of 0-elements specific for each class. The same holds for all the numbers. However the (0)::Complex is the same as the (0)::Real: they both refer to the 0-element of the class Field, instances of which are the classes Complex and Real.

The specificity of elements extends to further classes:

```
(4::Modulo[6])+(3::Modulo[6])
```

yields to

```
(1)::Modulo[6]
```

allowing the manipulation of Modulo variables while using the standard numbers. Relevant cases occur with objects of the type Modulo, 0-elements of domains, etc.

Automatic transformation of objects of one class into objects of a subclass whenever this is possible (for example 3 as PositiveInteger), is not supported in the current version of SYMO2, and the transformation has to be done manually. However due to the inheritance mechanism there is always the possibility of the transformation of an object of one class into an object of one of its superclasses:

```
In[8]:= b::NegativeInteger;
In[9]:= a+b
Out[9]= (3+b)::Integer
```

Another advanced characteristic partly supported by SYMO2 is the manipulation of intervals and through them the handling of arithmetic properties of functions. The usage of intervals for the addition of qualitative knowledge to functions and variables has been of major importance in Qualitative Physics [Bobrow 84] and Qualitative Simulation of functions [Quipers 86]. Recently it has been introduced in Symbolic Computation in the form of Properties [Weibel 91] and is implemented with the assume function in Maple [Gonnet 92]. In SYMO2 a variable can be declared as an element of an interval in the reals or integers:

```
In[10]:= x :: RealInterval[0,1]; y :: RealInterval[0,1];
In[11]:= x-y
Out[11]= (x-y)::RealInterval[-1,1]
In[12]:= x-x
Out[12]= (0)::Real
In[13]:= Sin[_?Real]::RealInterval[-1,1];
In[14]:= Sin[b] < 2
Out[14]= True
In[15]:= Sin[a]-Sin[b]
Out[15]:= (Sin[3]-Sin[b])::RealInterval[-2,2]
In[16]:= Sin[b]-Sin[b]
Out[16]= (0)::Real
```

The above examples show the expressive capabilities of SYMO2 in the manipulation of qualitative knowledge or properties in an abstract level with the usage of inheritance, declarations and intervals, without the need to build a theorem prover for this purpose.

6 Implementation

Implementation of the SYMO2 environment in the three examined functional languages is not homogeneous since they do not share common characteristics.

Extended usage of rules in MATHEMATICA is not supported by MAPLE and SCHEME and therefore rule emulation via methods had to be implemented for them. On the other hand there was a need to set a limit to the rule evaluation in MATHEMATICA and this was performed by introducing the method overriding technique. Multiple definitions of a function with the same name but different type of arguments within a program is not supported by MAPLE and dispatch techniques used in SCHEME had to be added. Nested evaluation of variables had to be emulated in MAPLE with the use of eval expressions. Inconsistencies in the evaluation of functions with side effects (set!, set-cdr! etc.) in SCHEME could not be left without changes. Problems of definition of new functions within the body of other functions in MATHEMATICA (MATHEMATICA 2.0 generates automatically new symbols in such cases) had to be overcome. As it is already discussed, MAPLE and MATHEMATICA do not provide an environment for parallel execution of functions, something that can be easily emulated in SCHEME.

For the implementation of SYMO2 a compromise between speed and functionality on the one side and system elegance and programming facilities on the other side had to be found. Error calls are limited to the minimum and inconsistent or undefined operations are simply ignored. This characteristic adds to the flexibly of SYMO2 and encourages the user to modify the working schema to a desirable extend. We hope that the results justify our choices.

7 Conclusions and Future Directions

In this paper parts of the SYMO2 programming environment with concern to symbolic computation applications were presented. Issues of Multiple Inheritance, Metaclassing and Instantiation have been discussed with particular emphasis on Algebraic Computation, with Domains and Polymorphic Operations.

SYMO2 in its current version allows the user of MAPLE, MATHEMATICA or SCHEME to develop quickly a first version (even incomplete) of the application by using classes as domains and it enables the incorporation of changes suggested by previous experiments. We have illustrated how these modifications are performed at the schema, method and object levels. We have described techniques which allow the fast execution of such modifications at run time, and their usefulness in computer algebra applications. Furthermore data encapsulation and concurrency issues are examined and their lack of support from algebraic computation systems as MATHEMATICA and MAPLE is discussed.

SYMO2 has solved the problems of schema modifications which appear in pure object oriented systems like SMALLTALK, and the integration of various programming paradigms into existing systems, without the need for creating one more interpreter. One of the main objectives is the studying of techniques which involve the use of data types and metaclasses for the modelling of algebraic problems and algorithms.

The improvements in the programming environment of the three examined systems were considerable with the addition of classes and hierarchy as this is described. Incremental polymorphic code can be used in MAPLE and SCHEME, rules can be safely ordered in various ways in MATHEMATICA with the usage of subclassing, relationships between various data types can be created and changed with the modification of inheritance links, etc. Furthermore the introduction of declarations and its usage as an operation for the the passing of various properties to symbols is a key issue in our work.

One of the future directions is the further experimentation with various programming paradigms and finally the construction of a complete consistent programming environment, based on the obtained knowledge, which is expected to override the weaknesses of the commonly used techniques by introducing features as the ones examined in this paper. The main objective is the maintenance of the consistency in the implementation of algebraic concepts by giving more value to the importance of the semantics and their mathematical context, an issue which has been very rarely addressed from the existing systems.

8 Acknowledgements

I thank J. Davenport, G. Gonnet, R. Maeder, S. Missura and M. Monagan for comments and improvements during the steps of conceptualisation, design and implementation of SYMO2. Experience has been gained during the preparation of material for courses on functional languages at ETH Zurich, in conversations with users of AXIOM, MAPLE, MATHEMATICA, ML and SCHEME and in argumentations with subscribers of various news groups.

References

[Abdali 86] S. K. Abdali, G. W. Cherry, N. Soiffer. An Object Oriented Approach to Algebra System Design. Proc. Of the 1986 Symp. on Symbolic and Algebraic Computation, ACM, 1986

[Banerjee 87] J. Banerjee (et al). Data Model Issues for Object Oriented Applications, ACM Trans. Office Info. Syst. 5, January 1987.

[Bradford 92] R. Bradford. Simplification of Multiple-valued Expressions. DISCO 92. Bath. UK. 1992.

[Coiffet 90] P. Coiffet, J. Zhao, F. Wolinski, P. Novikoff, D. Schmit. About Qualitative Robot Control, NATO Workshop on Expert Systems and Robotics, 1990.

[Cox 86] B. Cox. Object Oriented Programming. Addison-Wesley, 1986.

[Davenport 90] J. Davenport, B. Trager. Scratchpad's View of Algebra I: Basic Commutative Algebra. DISCO 90. Capri, Italy. Springer Verlag. 1990.

[Deutsch 84] P. Deutsch, A. Schiffman. Efficient Implementation of the Smalltalk-80 System. 11th ACM Symposium on Principles of Programming Languages, Salt Lake City, UT, 1984.

[Fateman 90] R. Fateman. Advances and Trends in the Design and Construction of Algebraic Manipulation Systems. Proc. of the ISSAC 90, ACM, 1990.

[Fateman 91] R. Fateman. A Review of Mathematica. (to be published in Lisp and Symbolic Computation). Revised version Sep. 16, 1991.

[Fortenbacher 90] A. Fortenbacher. Efficient Type Inference and Coercion in Computer Algebra. DISCO90. Capri, Italy. Springer Verlag. 1990.

[Geddes 90] B. W. Char, K. O. Geddes, G. H. Gonnet, M. B. Monagan, S. M. Watt. Maple Reference Manual, 5th edition, Waterloo Maple Publishing, Waterloo, Ontario, Canada, 1990.

[Goldberg 89] A. Goldberg. Smalltalk-80. second edition, Addison-Wesley, 1989.

[Gonnet 92] T. Weibel, G. Gonnet. An assume facility for CAS with a simple Implementation for Maple. DISCO92. Bath, UK, 1992.

[Limognelli 90] C. Limognelli, M Mele, M Regio, M. Temperini. Abstract Specification of Mathematical Structures and Methods. DISCO90. Capri, Italy. Springer Verlag. 1990.

[Maeder 91] R. Maeder. Programming in Mathematica. Addison-Wesley, second edition, 1991

[Mathlab 85] The Mathlab Group. Macsyma Reference Manual, MIT Press, 1985.

[Milner 90] R. Milner, M. Tofte, and R. Har. The definition of Standard ML, MIT Press, 1990.

[Moon 89] D. A. Moon. The Common Lisp Object Oriented programming Language standard. Appeared in Object-Oriented Cocepts, Databases, and Applications. W. Kim, F. H. Lochovsky ed. Frontier Series, ACM Press, 1989

[Neusius 91] C. Neusius. Synchronizing Actions. Proc. of the European Conference on Object-Oriented Programming 1991, Springer-Verlag, 1991.

[Niestrasz 89] O. Nierstrasz, D. C. Tsichritzis. Integrated Office Systems. Appeared in Object-Oriented Cocepts, Databases, and Applications. W. Kim, F. H. Lochovsky ed. Frontier Series, ACM Press, 1989.

[Jenks 85] R. Jenks, B. Trager, S. M. Watt, R S. Sutor. Scratchpad II Programming Language Manual, IBM, 1985.

[Peano 97] G. Peano. Logique Mathematique, from vol II of the Formulaire. 1897.

[Russell 10] B. Russell, A. N. Whitehead. Principia Mathematica, Cambridge University Press, 2nd edition, 1927.

[Santas 92] P. Santas. Multiple Subclassing and Subtyping for Symbolic Computation. Workshop on Multiple Inheritance and Multiple Subtyping. ECOOP 92. Utrecht, Netherlands. June 1992.

[Steele 84] G. L. Steele Jr. Common LISP: The Language, Digital Press, 1984.

[Sussman 85] H. Abelson, G.J.Sussman. Structure and implementation of Computer Programms. MIT Press, Cambridge, Mass., 1985.

[Tomlinson 89] C. Tomlinson, M. Scheevel. Concurrent Object-Oriented Programming Languages. Object-Oriented Cocepts, Databases, and Applications. W. Kim, F. H. Lochovsky ed. Frontier Series, ACM Press, 1989.

[Turner 85] D.A. Turner. Miranda: A Non Strict Functional Language with Polymorphic Types". Functional Programming Languages and Computer Architecture, Springer Verlag, 1985.

[Ungar 87] D. Ungar, R. B. Smith. Self: The Power of Simplicity. Proc. of the second ACM Conf. on Object Oriented Programming Systems, Languages and Applications, SIGPLAN Notices, vol. 22, 1987.

[Weibel 91] T. Weibel, G. Gonnet. An Algebra of Properties, ISSAC 91, ACM, 1991.

[Wolfram 88] S. Wolfram. Mathematica: A System for Doing Mathematics by Computer. Addison-Wesley,1988.

Building a Computer Algebra Environment by Composition of Collaborative Tools

Norbert Kajler*

kajler@sophia.inria.fr
Projet Safir, INRIA Sophia-Antipolis
BP 93, 06902 Sophia-Antipolis Cedex, France

Abstract. Building a software environment for Computer Algebra is quite a complex issue. Such an environment may include one or more Symbolic Computation tools, some devices, such as plotting engines or code generators, and a way to link others scientific applications. It is also expected that any of these components may be run on a remote processor and that the whole system is used via a convenient graphical user interface. The natural extensibility of Computer Algebra software, as well as the diversity of the needs expressed by their users, necessitate a highly open and customizable software architecture allowing different kinds of extensions and adaptations. Our approach consists of building the environment by composition of separately developed packages, using state of the art software engineering technologies in the spirit of the tool integration paradigm. This way, the different software components should be able to exchange data and freely cooperate with each other, without being too tightly coupled as in a monolithic system. A prototype of such an environment is currently under development in the framework of the SAFIR project. It will be built using an implementation of the software bus concept developed for the next version of Centaur, and should include a set of components, developed both internally and externally, and a homogeneous user interface.

Keywords: Computer Algebra, Software Bus, Tool Integration, User Interface.

1 Introduction

There is a clear need for a general purpose scientific software environment integrating various tools needed in mathematical and physical research, education and engineering activities. To solve a scientific problem, a typical user has to go through several steps: first, develop a formal description of the problem suitable for automated processing ; second, compute the solution using numeric and/or symbolic methods ; third, extract some relevant information from this solution and present it in a pleasant manner ; fourth, edit a document including the result in a more precise form and the description of the process needed to reach it. A non exhaustive list of possible software which could be used is: computer algebra systems, numerical libraries, graphical plotting engines, and scientific text processing tools. The aim of

* Also: I3S, 250 rue A. Einstein, 06560 Valbonne, France.

such an environment is to ease interactions and exchanges between these different tools and to gather them in a coherent manner for the casual or the advanced user.

Of course, each of these services is generally available as a separate program or library. Furthermore, much effort is being invested to develop these different programs. But, up to now, no existing system provides such a set of functionalities in a homogeneous and open application which also allows environment designers as well as experimented users to add or replace tools. That is why an interesting approach seems to be the development of an open architecture allowing both new and existing scientific tools to be integrated in a computer algebra software environment. Such an environment should also offer a high quality User Interface (in the sequel: UI) providing what is often missing from existing Computer Algebra Systems (in the sequel: CAS), that is: a way to cooperate with other applications and a modern graphical user interface.

2 The User Interface Problem

Most of the UI for CAS were designed for alphanumeric terminals. Even on fast graphic workstations, allowing 3D animated curves plotting, user interaction with the CAS engine is generally exclusively achieved via a command language which is used both for human-computer interaction and algebraic program writing. Also, important issues such as extensibility or communication with other scientific tools are usually not well supported by the UI.

By comparison, the UI of the best designed programs available on personal computers or Unix workstations are based on modern interaction paradigms such as direct manipulation or visual languages [3]. They were built after a careful study of the users' needs, using state of the art technologies in man-machine interaction. Besides, success of commercial software is more and more often due to the quality of their interfaces, which includes: overall ergonomy, compliance to industry standards, etc. Also, there is much pressure on software publishers to provide open programs ; we see that more and more packages are both highly customizable and designed to communicate with other applications.

In a previous paper [18], existing UI for CAS were classified and a set of possible specifications for a better UI were proposed and analysed. This first step of our work convinced us that the diversity of the CAS users as well as the variety of the CAS applications necessitate the building of a modular and extensible UI, making future adaptations less costly for the UI developer, and making also customizations possible for the different classes of users. Further investigations were then performed to specify what kind of extensibility was needed and how existing technologies could help us in building a convenient software framework. Our goal, now, is to develop a software architecture in which not only the UI, but also a more complete CAS environment, may be built by composition of different tools.

About the UI, two problems are generally not well addressed by existing systems, MathStation [22] excepted, that is: the extensibility of both the formula editor and the dialog manager. In our opinion, a full extensibility of the UI is a key problem in building a quality software environment around computer algebra, an important goal being to provide a convenient UI which could be adapted to the different application domains such as: Arithmetic, Robotics, etc, including their own symbols, graphic notations, and algorithm libraries.

- Dialog manager extensibility
  Whereas the CAS is generally written by a given team at a given location, library extensions may be written by any advanced user. In spite of this, it should be nice to have an homogeneous UI for both the CAS kernel and the packages written by the users. In fact, the UI, as well as the documentation, should tend to be as extensible as the CAS itself, providing to the user some kind of interactive algorithms driving. For instance, if a package implements an algorithm with certain variables and functions written in the CAS programming language, then loading this package should pop up a dedicated panel designed by the author of the package to provide a more convenient access to the new functionalities. Obviously, the system should help the package writers to build these graphical panels.
- Formula editor extensibility
  Using its programming language, the mathematical knowledge of a CAS may be easily extended by defining new operators. Also, by adding a set of rules for the CAS simplifier, it is generally possible to define different algebraic properties for these new operators. In the same manner, the formula editor should be extensible by defining new graphical operators and their associated formatting rules. However, with most existing scientific editors, this can be achieved only by the author of the program and entails re-compilation of the editor. So, the next step is to allow dynamic loading of specifications written by advanced users.

3 The Integration/Communication Problem

A classical integrated CAS such as Axiom, Derive, Maple, Mathematica, or Reduce is made of an algebraic part, which includes the kernel and the libraries, and a set of standard interfaces allowing, above all, the dialogue with the user, drawing curves in two or three dimensions, generating Fortran or LaTeX code, etc. Generally, these systems are delivered as a single program bringing together all these functionalities.

Extending these systems in terms of algebraic capabilities is quite natural: it involves programming algorithms in the CAS language. In this case, the problem is intimately linked to the algorithm complexity and to the optimizations to be brought, in terms of memory management for instance. On the other hand, the addition of new interfaces is generally hard work, even in the case of a simple extension. Very often, no high level mechanism is provided for aiding its implementation. However in the most modern systems, the CAS includes a communication protocol and an application programming interface allowing the communications between the CAS and an external program. This is the case of Maple V which proposes the Iris protocol [20] and of Mathematica with MathLink [32].

The limit of this approach is reached when the point is to create a complex environment in which every component may cooperate with all the others and not only with a given CAS. Also, to get a truly open system, it is necessary to allow runtime reconfigurations, which implies that every component should make no assumptions about which other components are present in the environment. More, the UI should offer an homogeneous and convenient way to interact with the complete set of functionalities provided by the system, which include the tools dynamically added in the environment and the extensions loaded by the components. Very few results have been reached on these topics, with the exception of some experimental systems such as Minion [25] or SUI [8].

When thinking in terms of software engineering, the integrated approach becomes a problem as soon as the complexity of the different elements calls for a separate development. this is precisely the case for a CAS kernel, the plotting engine, and the graphical formula editor. Reasonable estimations of the development cost for each of these packages may rank from three man years for a quality graphic formula editor, to fifty man years for a powerful CAS with a significant library. For these components, the cost of development justifies the implementation of separate entities or the re-using of possibly existent packages. On the same topic, the development of each package may become the speciality of specialized teams working independently. For instance, building a powerful plotting engine could be achieved by researchers interested in signal processing, while a mathematical formula editor could be produced by a team of specialists in scientific document processing. Also, the Computer Algebra part of the environment could result of the combination of complementary tools developed separately and specialized in some particular domains.

4 An Open and Distributed Architecture

Our objective is to build a scientific software environment in which the different functionalities that the user needs will be provided by independent software tools pluggable in the framework of a software architecture. In this context, we define a tool as being a software component providing a set of services that show strong internal cohesion and low external coupling [33]. Examples of such tools to be found in a scientific environment are a CAS, a formula editor, a code generator, etc. Further decomposition may be worthwhile: a larger tool or a toolset being considered as a composition of more elementary tools according to similar cohesion vs coupling criteria.

4.1 Architecture Openness

The software architecture will have to provide the "glue" needed to make the different components work together efficiently, without coupling the components too tightly as in a monolithic system. The desired degree of openness being that a tool may be added, deleted, or replaced with minimal perturbations to the rest of the environment. This will ease both the maintenance of the application and development of future extensions by the environment designers, as well as the addition of new functionalities by the users.

Also, as we anticipate these operations to be possible statically as well as dynamically, the UI could include a specific tool to ease runtime reconfiguration of the environment, possibly with the assistance of a visual language in the spirit of Cantata [14], the visual language used in Khoros to enable final users in building software treatment sequences by drawing graphs made of instances of the elementary software tools provided by the system.

4.2 Custom-made Scientific Environments

Another important possibility would be to develop, from a common set of components, some custom-made scientific environments, that is, systems specifically adapted to the needs of a particular class of users, or optimised to meet software constraints or hardware limitations.

In the first case, the selection and adaptation of the different components will be directed towards providing the features needed by the users. For instance, a scientific environment designed for educational needs may include a CAS with a simple library, a user-friendly UI, a 2D plotting engine, and a large set of tools and tutorials to help in using the system and understanding Mathematics, Mechanics, etc. In contrast, High Energy Physics specialists may need a different environment with a more powerful CAS able to manage very large expressions, a tool to generate Fortran programs, and a graphical UI including a formula editor tailored to display specific notations.

In the second case, the components will be selected by the environment designer with the aim of controlling precisely the resources needed for the execution, such as main memory usage or hardware requirements. This way, small systems including a CAS with its minimal library and a textual UI may be delivered as well as more sophisticated environment including the CAS, a multi-window graphic UI and a more complete set of capabilities.

4.3 Distributed Environment

Finally, we anticipate each component being a separate process running on different nodes of a local network. Distributed computation will be supported transparently by the software bus architecture which should provide a set of high level communication mechanisms such as send or notify, built on top of lower level routines such as RPC. This way, it will be easier to run the different tools on different processors in order to take full advantage of specialized hardware optimised for numerical, symbolic, or graphical work.

5 Technologies Involved

The tool integration concept consists in building an application as a set of co-operating components. This principle is progressively implemented within the frame of programming environment projects such as Arcadia [30], Centaur [17], or HP Soft-Bench [6], and seems to be a nice model for the architecture of a complex computer algebra environment.

Quoting Schefström [28], tool integration is the sum of three levels of integration: data, control, and user interface. Together, they make possible an efficient co-operation between the components brought together in an homogeneous and highly open system.

5.1 Data Integration

Data integration is related to the sharing and exchange of data between the different components. It determines the degree to which data generated by one tool is made accessible and is understood by other tools.

A general solution is to define common abstract models which each component may interpret [31]. Here, a portable encoding protocol for mathematical expressions will be a first step in this direction. This will ease data exchange between different kinds of applications running on the nodes of an heterogeneous local area network. Unfortunately, this method does not address the data sharing aspect of data integration which includes some important issues such as data duplication, constancy, and persistence.

Also, sophisticated methods for data abstraction derived from the Portable Common Tool Environment (PCTE) [4] or the Interface Description Language (IDL) [29] could be investigated in the future. This could lead to the creation of an algebraic data server under the form of an active node used by the other software components as a formula repository. Two levels of use are envisaged for the formula server. First, components with their own internal data representation would notify and query the formula server using a coercion protocol between the component data and the server data. Second, some components may use the data server as a transparent set of high level algebraic data structures they could access via a predefined interface. In this case, no coercion should be needed, but the tool should connect often to the server and could not manage directly its data. In both cases, the server will centralize data and make them available for the different components using a uniform data access interface providing navigation, modification, and referencing primitives. Of course, the data server will ensure both data coherence, needed when more than one component sends a destructive modification request on a same data instance; and data persistence, making data existence possibly independent of the life of the component which had created them.

Different kinds of solutions are possible here, however, we may be aware of some Computer Algebra characteristics such as the use of slightly complex data structures and the need to control precisely these data structures in order to write efficient algorithms. Also, it would not be reasonable to adopt a solution which could not be adapted to both new and existing tools.

5.2 Control Integration

Control integration concerns inter-tools communications. It determines the degree to which each tool inter-operate with the others, i.e communicate its findings and actions to other tools and provide means for other tools to communicate with it. In fact, inter-tool communication complements data integration and is a very general issue within software systems. Recent works, such as hypertexts [5] and extension languages in the spirit Tcl [24], indicate that a good approach is to leave communication control to be external to the tool. This generally leads to more powerful systems in which inter-objects links are used to exchange data or activate external routines transparently.

Using this approach as the basis of our computer algebra environment, we plan to leave the coupling of the different tools to advanced users or environment designers.

This could be achieved either by writing some scripts to specify the connections and message routing, or by drawing a task scheme using a dedicated component of the user interface. A challenging goal is to allow a final user to set up some complex processing tasks involving different tools such as a formula editor which send an expression to a CAS which returns another expression displayed in the editor and simultaneously dispatched to a numerical application which will produce a set of points visualized by a plotting engine.

This may be achieved either by tools explicitly addressing each other, or by means of a a typed messages broadcasting mechanism in the spirit of Field [27]. The second solution is obviously the most desirable as it largely eases extensibility and dynamic reconfigurations. Also, the Field approach is used in most of the "software bus" implementations which are emerging now, including Hp SoftBench [6] or Sun ToolTalk [16].

5.3 User Interface Integration

Finally, the aim of user interface integration is to ensure the same style of interaction for the whole system. A homogeneous UI should hide the granularity of the implementation and reduce user's cognitive load. This include using a uniform and standard look and feel, and also factorizing similar metaphors and mental models used in the different parts of the user interface.

This may be achieved by a decomposition of the UI into two parts: the external user interface which will manage usual interactors such as buttons, menus, panels, etc, and the internal user interface which corresponds to the set of specific editors needed by the application.

For the external user interface, an interesting solution consists to encapsulate a popular toolkit providing a standardized look and feel, such as Motif [23], under the form of a widget server, and to use this server for any dialog with the user. Ideally, the global user interface should not merely be the superposition of the user interfaces of the different components. On the contrary, it should factorize as much as possible the similar dialog mechanisms (selectors, control panels, etc). For this purpose, another component, named dialog server, could centralize every interactions with the user, and doing so, improve the global ergonomy by reducing the number of interactors on the screen.

For the internal user interface, we need a different class of toolkit to help display application specific data, that is mathematical objects such as formulas, three dimensional curves, commutative diagrams, etc. Such a toolkit should provide the necessary primitives to display very different kinds of graphics and to allow mouse based referencing of structured subparts in the corresponding application data. The definition of a general graphic toolkit is still an open question and many efforts are still made to get effective results on this subject. This includes, higher level graphic toolkits such as Interviews [21] or Andrew [1], document processing toolkits or editorial workbenches such as Framemaker or Grif, and interactive programming environments such as Centaur [17] or The Visual Programming Workbench [13]. For instance, from now on, one may derive a graphical structured editor from a set of specifications using attributed grammars like in Gigas [11], or using some specialized editor definition languages like in Grif [26].

6 Implementation of a Prototype

A prototype of this environment, named CAS/π[2] [19], is currently being developed in the SAFIR project at INRIA and University of Nice-Sophia Antipolis. The goal is to get quickly a first operational release on which further experiments may be realized in the framework of future researches on scientific software environments and related communication protocols.

Our environment prototype will gather four kinds of components: CAS(s), curves plotter, graphical formula editor, and dialog manager. For the first version of the prototype, we will experiment on Sisyphe [12] and Ulysse [9], the two experimental Computer Algebra engines developed locally by the SAFIR project, and ZIClib [10], the surface mesh visualization library. The formula editor and the dialog manager will be generated with the help of software tools, in order to ease future modifications and extensions. For instance, we are experimenting with the incremental formatter Figue [15] to generate a formula editor and a set of formula browsers; and we plan to build our dialog manager on a Motif server [2] developed for the future version of Centaur [17]. Progressively, we will try to improve the ergonomy of these two components to satisfy the needs expressed by different classes of CAS users.

Also from the next version of Centaur, we are using an implementation in LeLisp of the control integration concept, based on asynchronous typed message broadcasting [7]. According to this implementation, each tool will be encapsulated under the form of an integrated software component and will communicate with the rest of the environment by sending and receiving typed messages. The principle of the broadcasting being that a message emitted by a component is not directly addressed to another component named in the message header, but rather sent out and dispatched to every component registered as a recipient of that type of message. Obviously, this message-routing mechanism is managed in a transparent way by the tool integration technology, providing a "software bus" basis layer for the rest of the environment.

7 Conclusion

Recent progress in Software Engineering as well as the coming out of new technologies from the Unix world make possible the development of highly open and distributed software architectures. In the field of computer algebra, it seems interesting to apply these concepts to design a scientific software environment enabling to gather numerical, visualization, scientific document processing, and symbolic computation tools.

Improving the user interface, which was historically at the beginning of these developments, would also mean realizing a set of separate components providing a convenient graphical man-machine interface for the whole environment. This should include a sophisticated and extensible graphical formula editor, coupled to a dialog manager server. This may also include a visual language to allow dynamical reconfiguration of the inter-component interactions.

[2] We plan do demonstrate CAS/π in July 92 for the next ISSAC conference.

On the other side, the opening of the picked up architecture will provide a convenient basis to experiment alternative user interface approaches or inter-tool communication protocols. Obviously, an open and extensible software architecture will ease the integration of further technological progress in the field of man-machine interaction, scientific document processing, or data integration. A key motivation for us being our conviction that more and more complex applications will be available not only as stand alone monolithic programs, but also under the form of toolkits designed to ease a selective re-using.

Acknowledgements

The author would like to thank Marc Gaëtano, André Galligo, and José Grimm of the SAFIR Project, Dominique Clément of the SEMA group, Jack Wileden of University of Massachusetts, and the CROAP project led by Gilles Kahn for their great help right from the beginning of this work.

References

1. Palay J. Andrew. The Andrew Toolkit: An overview. In *1988 Winter USENIX Technical Conference*, pages 9–21, Dallas, Texas, February 1988.
2. Ali Atie. *Third annual GIPE report*, chapter LLMI: An Interface Between LeLisp and OSF/Motif. ESPRIT Project 2177, 1991.
3. L. Bass and J. Coutaz. *Developing Software for the User Interface.* Addison-Wesley, 1991.
4. G. Boudier, F. Gallo, R. Monot, and R. Thomas. An Overview of PCTE and PCTE+. In *ACM Software Engineering Symposium on Practical Software Development Environments, SIGSOFT Software Engineering Notes, 13(5)*, November 1988.
5. CACM. Special issue on hypertext. *Communications of the ACM*, July 1988.
6. M. Cagan. HP Soft Bench: An Architecure for a New Generation of Software Tools. SoftBench Technical Note Series SESD-89-24, Hewlett-Packard, November 1989.
7. Dominique Clement. A distributed architecture for programming environments. Rapport de recherche 1266, INRIA, July 1990.
8. Y. Doleh and P. S. Wang. SUI: A system Independent User Interface for an Integrated Scientific Computing Environment. In *ACM Proc. of the International Syposium on Symbolic and Algebraic Computation*, pages 88–94, Tokyo, August 1990. Addison-Wesley.
9. C. Faure, A. Galligo, J. Grimm, and L. Pottier. The extensions of the Sisyphe computer algebra system: Ulysse and Athena. In *Proc. of DISCO'92*, Bath, GB, April 1992. Springer-Verlag.
10. R. Fournier. *ZICVIS et la ZIClib.* Rapport interne INRIA, Sophia-Antipolis, April 1992.
11. Paul Franchi-Zannettacci. Attribute Specifications for Graphical Interface Generation. In *IFIP 11th World Computer Congress*, pages 149–155, San Francisco, USA, 1989.
12. André Galligo, José Grimm, and Loïc Pottier. The design of SISYPHE : a system for doing symbolic and algebraic computations. In A. Miola, editor, *LNCS 429 DISCO'90*, pages 30–39, Capri, Italy, Avril 1990. Springer-Verlag.
13. E. Golin, R. V. Rubin, and J. Walker II. The Visual Programming Workbench. In *IFIP 11th World Computer Congress*, pages 143–148, San Francisco, 1989.

14. The Khoros Group. *Khoros Manual, Release 1.0.* University of New Mexico, Albuquerque, 1991.
15. Laurent Hascoët. *FIGUE: An Incremental Graphic Formatter. User's Manual for Version 1.* INRIA, Sophia-Antipolis, August 1991.
16. SunSoft Inc. The ToolTalk Service. Technical report, September 1991.
17. Gilles Kahn et al. CENTAUR: the system. In E. Brinksma, G. Scollo, and C. Vissers, editors, *Proc. of 9th IFIP WG6.1.* Intern. Symp. on Protocol Specification, Testing and Verification, 1989.
18. Norbert Kajler. Building Graphic User Interfaces for Computer Algebra Systems. In *Proc. of DISCO'90*, pages 235–244, Capri, Italy, April 1990. LNCS 429, Springer-Verlag.
19. Norbert Kajler. CAS/PI: a Portable and Extensible Interface for Computer Algebra Systems. In *Proc. of ISSAC'92*, Berkeley, USA, July 1992. ACM Press. To appear.
20. B. L. Leong. Iris: Design of a User Interface Program for Symbolic Algebra. In *Proc. of the 1986 ACM-SIGSAM Symp. on Symbolic and Algebraic Computation*, July 1986.
21. Mark. A. Linton, Paul. R. Calder, and John. M. Vlissides. InterViews: A C++ graphical interface toolkit. Technical Report CSL-TR-88-358, Stanford University, July 1988.
22. MathSoft, Inc., One Kendall Square, Cambridge, MA. *MathStation, Version 1.0.*, 1989.
23. OSF/Motif. *OSF/Motif Programmer's Guide, Programmer's Reference Manual & Style Guide*, Open Software Foundation edition, 1990.
24. J. K. Ousterhout. Tcl: An embeddable command language. In *1990 Winter USENIX Conference Proceedings.* Univ. of California at Berkeley, 1990.
25. James Purtilo. Minion: An Environment to Organize Mathematical Problem Solving. In *Symposium on Symbolic and Algebraic Computation*, pages 147–154. ACM, July 1989.
26. Vincent Quint. *Structured documents*, chapter Systems for the representation of structured documents, pages 39–73. The Cambridge Series on Electronic Publishing. Cambridge University Press, Cambridge, 1989.
27. Steven P. Reiss. Connecting Tools Using Message Passing in the Field Environment. *IEEE Software*, pages 57–67, July 1987.
28. Dick Schefström. Building a Highly Integrated Development Environment Using Preexisting Parts. In *IFIP 11th World Computer Congress*, San Francisco, August 1989.
29. Richard Snodgrass. *The Interface Description Language: Definition and Use.* Principles of Computer Sciences Series. Computer Science Press, 1991.
30. R. Taylor, F. Belz, L. Clarke, L. Osterweil, R. Selby, J. Wileden, A. Wolf, and M. Young. Foundations for the Arcadia Environment Architecture. In *ACM SIGSOFT'88: Third Symposium on Software Development Environments*, November 1988.
31. Jack C. Wileden, Alexander L. Wolf, William R. Rosenblatt, and Peri L. Tarr. Specification Level Interoperability. In *12th International Conference on Software Engineering*, pages 74–85. IEEE Computer Society Press, May 1990.
32. Wolfram Research, Inc. *MathLink External Communication in Mathematica*, 1990.
33. E. Yourdon and L. L. Constantine. *Structured Design: Fundamentals of a Discipline of Computer and System Design.* Prentice-Hall, 1979.

An assume facility for CAS, with a sample implementation for Maple

Trudy Weibel
School of Mathematics and Statistics
University of Sydney NSW 2006, Australia

Gaston H. Gonnet
Institute for Scientific Computing
ETH, 8092 Zürich, Switzerland

Abstract

We present here an assume facility for computer algebra systems. The tool is to facilitate a means to resolve branching problems in a computer algebra program of the kind "`if object has a certain property then` ... else ...". The approach is based on the key idea that first the calculations performed on an object is duplicated on the object's initial property. The so computed property can then be tested against the primary queried property. To demonstrate that this concept in realisable we implemented it in Maple.

1 Introduction

Traditional computer programs basically deal with numerical values. They return numbers as results which then may be interpreted say as coordinates for a plot. In contrast, computer algebra programs deal, principally, with symbolic objects,i.e. variables may remain unassigned. Over the last decade such computer algebra systems (CAS) grew rapidly in strength in what we call calculus. But now, they touch a qualitative limit of a kind that does not occur in numerical programming. Let us look at the example

$$\int_0^\infty xe^{-ax}dx = \begin{cases} \frac{1}{a^2} & \text{if } a > 0 \\ \infty & \text{otherwise}. \end{cases}$$

In a conventional program, the variable a will be assigned some value. Thus the if-guard "> 0" can be tested and subsequently it can be decided which branch to chose. However, if there are unassigned variables involved, as this is characteristically the case in computer algebra programs, we need some new means to determine the if-guard. A typical situation is as follows. We start off with an unassigned real variable b. During the course of a computer algebra program, the variable a is assigned the square of b and then appears in the above integral. Any calculus student would now know which branch of the integral to choose, but how does a CAS know which path to proceed? This is the type of problem that our proposed assume facility for CAS is to solve.

Essentially, we regard such decisions as queries on objects for conditions to hold (Section 2). Behind such a condition lies the concept of a property. This builds the framework within which we can handle such decisions in a uniform manner. In fact, properties of objects constitute an algebra **PROP**. Together with some additional axioms a calculus of properties is established (Section 3). In order to determine the query during a running program the necessary information needs to be made available. The tool we propose is an assume facility that allows the user to make assumptions on otherwise undetermined objects. It has been implemented for a CAS, Maple (Section 4).

It should be pointed out here, that this class of problems cannot be addressed with a typing method. The above example makes it clear that it is not a matter of allowing or forbidding certain operations on the grounds of formal, syntactical reasons, but rather a question of semantic bearing. A close investigation of the relationship between the theory of properties and type theory is forthcoming.

2 Reducing the query to a property problem

As instigated by the introductory example, we want, within a CAS, to be able to decide predicates such as "$a > 0$" under assumptions such as "$a = b^2$ and b is a variable for a real number". More generally, we assume a given set of objects to satisfy certain properties and then ask for the validity of some property of an object composed of the given ones:

$$\text{Given: } obj_1 \in prop_1, \ldots, obj_n \in prop_n$$
$$\text{Query: Is } f(obj_1, \ldots, obj_n) \in prop_0\,?$$

In order to determine this query, [WG91] propose to split the problem up into two parts. (Note that within this framework, properties are considered as sets of objects.)

1. We transform the properties along with the objects:

$$\begin{array}{ccc} (obj_1, \ldots, obj_n) & & (prop_1, \ldots, prop_n) \\ \downarrow & & \downarrow \\ f(obj_1, \ldots, obj_n) & & \overline{f}(prop_1, \ldots, prop_n) \end{array}$$

2. This reduces the query to the problem:

$$\text{“Is } \bar{f}(prop_1,\ldots,prop_n) \subseteq prop_0 \text{ ?”}$$

To obtain a correct result, we only need to insure that the transformation "bar" respects the incidence relationship "$\in$" ($obj \in prop \Rightarrow f(obj) \in \bar{f}(prop)$).

Thus, we first determine the transformed property, $\bar{f}(prop_1,\ldots,prop_n)$,—the harder and more involved part of two steps. Afterwards, we merely have to check whether the transformed property is contained in the queried property.

This approach shifts the heterogeneous query problem involving objects and properties to a pure property issue. In the next section, we present the model of this new world of properties, the algebra **PROP**. It is a lattice structure satisfying additional axioms, which leads to the rules for computing with properties.

3 The theoretical framework: the algebra PROP and its axioms

Here we only give a rough account of the theoretical framework underlying the concept of the proposed assume facility. Originally and in detail the theory on properties has been presented in [WG91].

Treating properties as sets of objects, the predominant structure that they embody is a lattice structure, in fact a boolean algebra. The lattice operators $\neg$, $\wedge$, and $\vee$ are appropriately induced by the ordering relation $\leq$ which is the correspondent to the interpreted set inclusion $\subseteq$.

$$\mathbf{PROP} := \langle \mathrm{Prop}_0, \neg, \wedge, \vee, \top, \bot \rangle \quad \text{a Boolean algebra,}$$

where the symbols are interpreted for properties as follows:

Prop_0	a given set of basic properties (e.g. "real", "positive", "to be a type")
$\leq$	"inclusion": the LHS property is included in the RHS property
$\neg$	"not", as in "not positive"
$\wedge$	"and", as in "integer and negative"
$\vee$	"or", as in "positive or negative"
$\top$	some property ("existent") that holds for *any* object
$\bot$	some property ("inexistent") that holds for *no* object

In the previous section we saw that in handling queries on composed objects, we need a corresponding mapping on the properties of the given objects. That is, for a given object function, f, we need a corresponding property function, $\bar{f}$. Hence, the model **PROP** supplies, besides the boolean operators, certain property function symbols. As it turns out, the model allows one to restrict the set of property function symbols to those that correspond to basic object functions, and inverses thereof. In particular, if we are to transform a composition of two functions, $f \circ g$, we only have to provide for $\bar{f}$ and $\bar{g}$ separately, since composition can be carried through the "bar"-transformation: $\overline{f \circ g} = \bar{f} \circ \bar{g}$. If we collect these basic property functions symbols in $\mathcal{F}$ the model for the properties is the following structure:

$$\mathbf{PROP} := \langle \mathrm{Prop}_0, \neg, \wedge, \vee, \top, \bot, \mathcal{F} \rangle,$$

i.e., as our properties, we have some basic, given properties such as "positive" or "negative" and composite properties such as "positive $\vee$ negative" or "$\overline{square}(negative)$".

So far, we only have a description of the properties, what they are composed of and how they are built from more basic ones. However, we are interested in computing such composite terms, i.e. what are the laws that hold among them. For a detailed presentation of the set of laws we have to refer again to the earlier paper [WG91]. Besides the classical lattice axioms the three key axioms that allows an effective calculus on properties are those of monotonicity and distributivity:

Monotonicity:	$a \leq b$	$\Rightarrow$	$\bar{f}(a) \leq \bar{f}(b)$
Distributivity over $\vee$:	$\bar{f}(a \vee b)$	$=$	$\bar{f}(a) \vee \bar{f}(b)$
Weak Distributivity over $\wedge$:	$\bar{f}(a \wedge b)$	$\leq$	$\bar{f}(a) \wedge \bar{f}(b)$
(if "$\bar{f}$ is injective on $a \vee b$":	$\bar{f}(a \wedge b)$	$=$	$\bar{f}(a) \wedge \bar{f}(b)$)

These constitute the rules according to which the computations on properties are carried out. In the next section we show how we incorporated them in a computer algebra system such that they rule an assume facility which handles our queries on properties of objects during a running symbolic algebra program.

4 The implementation for Maple

We opted for the CAS Maple to carry out an implementation of an assume facility. The choice was mainly based on Maple's strength in its programming language. It allows to implement a complex subsystem, such as the assume facility, entirely within the provided, very convenient, high level programming language. For an integral description of Maple we refer to the two manuals on its language and its library, [CGG+91a] and [CGG+91b].

4.1 Objects and properties

Dealing with queries or assumptions, there are two kinds of entities involved: On one hand we have the objects, which are the main objectives of the program. On the other hand we have the properties, to which objects refer to, and which are to be handled and computed by the program along with the objects.

In Maple, we can consider the valid expressions as our objects and specified subsets thereof as the possible properties, typically names (e.g. **rational**), types (e.g. **numeric**), real intervals (e.g. **RealRange(0,infinity)**). For reasons of demonstration, we have restricted the objects to expressions of type 'algebraic' and the properties chosen are those that relate to these algebraic objects.

4.2 The system

The assume facility presented here constitutes a subsystem of Maple, neither a system in its own merit nor a "Maple package". The facility maintains information (i.e. properties) on objects, which is partially initialized through the system and partially contributed by the user. Moreover it provides the user with some functions (e.g. **assume**, **is**) to supply or retrieve information.

Interface

The present implementation is based on the following interface:

- the function **assume**, which tells the system which are the new properties of an object to be known;
- properties of the objects are stored in the **ObjectTable**;
- the function **is**, which determines (computes) whether a given property is true, false or undecidable;
- the function **isgiven**, which just returns true or false depending on whether a given property has been specified (without attempting to do a complete derivation);
- the function **about**, which displays information to the user about the current status of objects or properties;
- the tables **ParentTable** and **ChildTable**, which define the immediate parents and children of properties in the property lattice;
- the function **addproperty**, which installs a property in the proper tables given their parents and children;
- the property functions **'property/<function>'**, which, given the property **<property>**, return a property **'property/<function>'(<property>)**, such that the following holds: if **x** has property **<property>**, then does **<function>(x)** have the property **'property/<function>'(<property>)**.

Representation

The main constituents of the assume facility are the properties. They are represented in our system in the following ways. A property can be

a)	a name	(has to be placed in the lattice) e.g. **rational, diagonal, TopProp**
b)	most types	(including constants) e.g. **integer, numeric, set, 0**
c)	a numerical range	e.g. **RealRange(-infinity,Open(0))**
d)	an "and" of properties	(corresponds to intersection) e.g. **AndProp**($prop_1$, $prop_2$, ...)
e)	an "or" of properties	(corresponds to union) e.g. **OrProp**($prop_1$, $prop_2$, ...)
f)	a property range	(an interval in the property lattice) e.g. **PropRange(Open(prime),rational)**
g)	a parametric property	(represented as unevaluated function calls) e.g. **BlockDiagonal(n,regular)**.

The interdependence of the basic properties (of the form a) and b)) can be visualized with a lattice diagram. Figures 1 and 2 show examples of two partial lattices that define the numerical and functional properties of the current

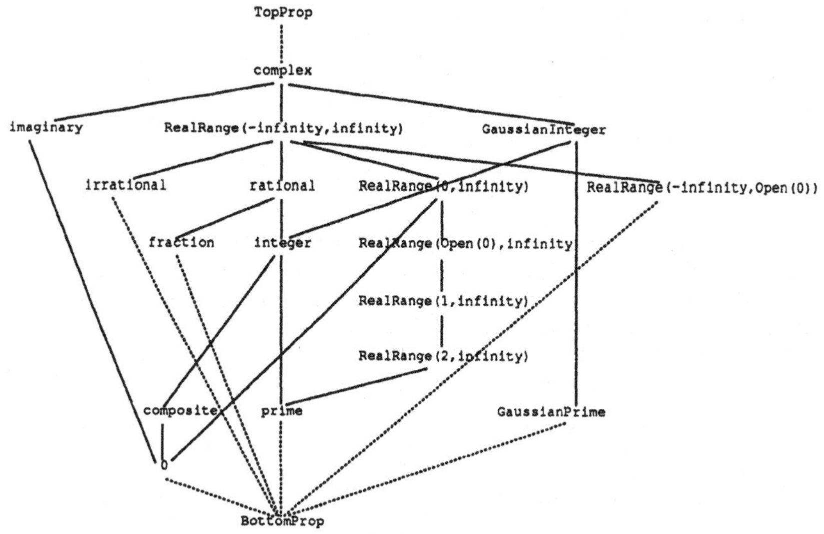

Figure 1: Lattice for numerical properties.

implementation. There, an arc from a lower to an upper property means that the former is contained in the latter.

Such a property lattice is represented in our implementation by two tables for the immediate ancestors and descendants, **ParentTable** and **ChildTable**. For example, **ParentTable[integer]** is assigned the set {**rational**, **GaussianInteger**}. While the parent table is explicitly preloaded with numerical, functional, matricial and other basic (like **TopProp**) properties (as indicated in Figures 1 and 2), the child table is created automatically at initializa-

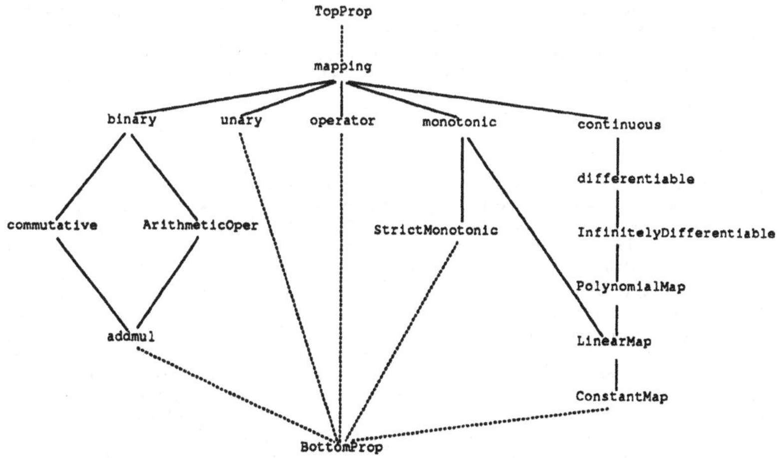

Figure 2: Lattice for functional properties.

tion time. Primarily, it is envisaged that the set of basic properties remains fixed. However, the user can add a new named property or a numerical range via an explicit call to **addproperty**, e.g., **addproperty(LinearMap,{monotonic,**

PolynomialMap},{ConstantMap}). The storage of all adjacency relations in the lattice allows such an update to be performed locally: addproperty affects only parent entries of all the children of the new property and child entries of all the parents of the new property, together with an initialization of the two tables for the new property. In above example, the update changes ParentTable[ConstantMap], ChildTable[monotonic], ChildTable[PolynomialMap] and initializes ParentTable[LinearMap] and ChildTable[LinearMap].

The remaining manifestations of properties, RealRange,
PropRange,
AndProp,
OrProp and
<parametric>
are handled as data structures for properties. They are implemented using an object oriented approach in Maple:

the OO data structure as an unevaluated function call.

E.g. AndProp(prime,GaussianPrime). Thus, polymorphic usage and variable arity is allowed for.

At the same time, the function calls perform the simplification (or normalization) particular to that data structure. For example, the function AndProp carries out following simplifications:

AndProp(a) $\Longrightarrow$ a
AndProp( ) $\Longrightarrow$ TopProp
AndProp(...,a,...,a,...) $\Longrightarrow$ AndProp(...,a,...)
AndProp(...,a,...,b,...) and a $\subseteq$ b $\Longrightarrow$ AndProp(...,b,...)
AndProp(...,AndProp(...),...) $\stackrel{\text{flatten}}{\Longrightarrow}$ AndProp(...,...,...)
AndProp(...,a,...,OrProp(...,a,...),...) $\stackrel{\text{absorption}}{\Longrightarrow}$ AndProp(...,a,...).

When no simplifications can be applied, the function returns unevaluated.

E.g., AndProp(GaussianInteger,rational,complex) $\stackrel{\text{simplify}}{\Longrightarrow}$ AndProp(integer) $\stackrel{\text{simplify}}{\Longrightarrow}$ integer;

AndProp(GaussianPrime,integer,integer) $\stackrel{\text{simplify}}{\Longrightarrow}$ AndProp(GaussianPrime,integer).

Note that generally, these composed properties are not explicitly included in the lattice diagram. Rather, their place in the lattice is computed on demand.

The properties as a whole constitute a type in Maple. 'type/property' tests an expression whether it fits one of above descriptions of a property.

While the representation of the properties play a pivotal role for the computations involving properties, our primary interest lies with associating object with properties. We represent the information what objects have what properties with a table, the ObjectTable, e.g. ObjectTable[cos] = AndProp(InfinitelyDifferentiable,unary). On one hand, the system preloads a big supply of such knowledge (its extent can be examined with a call to ObjectTable()). On the other hand, the user can store such information explicitly via the assume function (see also next paragraph), e.g. assume(x-1,RealRange(0,infinity)).

Implementation of the main functions

assume is the main function available to the user to provide the system with information, namely to declare which properties a certain object is to have. For example, assume(x, AndProp(rational,RealRange(0,infinity))) updates (or generates) a new entry in the ObjectTable (ObjectTable[x] := AndProp(rational,RealRange(0,infinity))). To accommodate a special, but very natural situation, the function assume (as well as is and isgiven) accepts, besides a pair of arguments, a single argument if it is of type relation.

When providing this assume functionality to the Maple system as such, a side problem arises in context with the 'remember' feature: Say the user wants to evaluate the formula int(exp(-a*t),t=0..infinity), where a is an unknown, and Maple returns unevaluated. The user may then assist the system by supplying some information about the unknown a, assume(a>0), and then let Maple evaluate the expression once more. Because of the 'remember' feature the system does not reevaluate the expression, but returns the previous answer, namely unevaluated. To overcome this dilemma without loosing the 'remember' functionality a reassign mechanism was introduced in the assume procedure. Now, when invoking assume(obj,prop), a new variable is assigned to at least one of the objects that obj is composed of. In our example, assume(a>0) has the side effect: a := a*. Thus, 'remember' recognizes the formula, now as int(exp(-(a*)*t),t=0..infinity), as a new one, one that has to be evaluated afresh.

In contrast to the "feed in" function assume, the two functions isgiven and is retrieve such information. The first function, isgiven, merely searches the ObjectTable for a direct derivation whether the queried object satisfies the queried property.

The second function, is, lies at the very heart of the assume facility. There, a query (i.e. is(object,property)) is "computed" by applying the lattice axioms and the property calculus rules (monotonicity and distributivity laws, cf.

Section 3). This function constitutes the core of the system wrt. both the expected crucial answer and the complexity that its computation involves. The way an answer is found is along the outlines of section 2:

- Query: — For example:
  `is(obj,prop)` — `is(x+2y>3)` (is reformulated to `is(x+2y-3,RealRange(Open(0),infinity))`)
- We assume that we know from the `ObjectTable`:
  $\texttt{obj}_1$ has $\texttt{prop}_1$ — `x-1` is `RealRange(Open(0),infinity)`
  $\texttt{obj}_2$ has $\texttt{prop}_2$ — `y-1` is `RealRange(0,infinity)`
  ⋮
- 1st step: `obj` is expressed in terms of $\texttt{obj}_i$:
  $\texttt{obj} = \texttt{f}_1(\texttt{obj}_1, \texttt{obj}_2, \ldots)$
  $\texttt{obj} = \texttt{f}_2(\texttt{obj}_1, \texttt{obj}_2, \ldots)$ — $\texttt{x+2y-3} = \texttt{obj}_1\texttt{+2obj}_2 = \texttt{(x-1)+2(y-1)}$
  ⋮
- 2nd step: Compute all the $\overline{\texttt{f}}_i(\texttt{prop}_1, \texttt{prop}_2, \ldots)$:
  `RealRange(0, infinity)` $\overline{+}\,\overline{2}$ `RealRange(0, infinity)`
  ⋮ ⇓
  `RealRange(Open(0), infinity)`
- 3rd step: Determine whether any of the $\overline{\texttt{f}}_i(\texttt{prop}_1, \texttt{prop}_2, \ldots)$ is included in prop:
  $\texttt{PropInclusion}(\overline{\texttt{f}}_i(\texttt{prop}_1, \texttt{prop}_2, \ldots), \texttt{prop})$
  ⋮ `RealRange(Open(0), infinity)` $\subseteq$ `RealRange(Open(0), infinity)`

Step 2 contains the intricate computation of $\overline{\texttt{f}}_i(\texttt{prop}_1, \texttt{prop}_2, \ldots)$. A general, possibly composed, function `f` is treated in the system intern procedure `Function`, which computes either an exact value or upper and lower limits (i.e. returns a `PropRange` as its value). It in turn is supported, again system intern, by explicitly given basic property functions, such as `'property/+'` (its Maple code is given as an example in Appendix B).

For more details, the complete Maple code for the function `is` is given in Appendix A.

The last of the main user functions, about, has no functionality, but is basically an information tool. It yields a verbose description of properties, of objects and their associated properties, as well as other information used by the assume facility. It is intended to be read by the user similar to a help file. For example,

```
about(real)
real:
  property aliased to RealRange(-infinity,infinity)
  a real range, in mathematical notation: (-infinity,infinity)
  a known property having {complex} as immediate parents
    and {rational, irrational, RealRange(-infinity,Open(0)), RealRange(0,infinity)}
    as immediate children.
```

Below is a complete list of all the data structures and procedures pertaining to the assume facility.

Data structures:	`RealRange`	`AndProp`	
	`PropRange`	`OrProp`	
Tables:	`ParentTable`	`Description`	
	`ChildTable`	`ExclusiveTable`	
	`ObjectTable`		
Sets:	`NumeralNames`	`basicPropertyFunctions`	
	`FunctionalNames`	`Aliased`	
	`MatricialNames`	`Aliases`	
	`OtherNames`		
Procedures:	`assume`	`ParentClosure`	`ConvertRelation`
	`is`	`ChildClosure`	`Exclusive`
	`isgiven`	`IncludedParents`	`AreExclusive`
	`about`	`IncludedChildren`	`LLLess`
	`addproperty`	`PropInclusion`	`RLLess`
	`'type/property'`	`LatticeInclusion`	`RRLess`
	`'type/AndProp'`	`RealInclusion`	`Function`
	`'type/OrProp'`	`NumericInclusion`	`'property/+'`
	`'type/PropRange'`	`AndOrInclusion`	`'property/*'`
	`'type/RealRange'`	`ObjProperty`	`'property/^'`
	`'type/ParamProp'`	`UpperBounded`	`'property/abs'`
	`VerifyProperty`	`WithProp`	
	`VerifyIntegrity`	`EraseRemember`	

When the assume facility is invoked in a Maple session, it first initializes the property lattice, some explicit values for property functions, possibly some ObjectTable entries and checks the integrity of the whole initial state of properties. Then the system is ready to digest new information and reply to requests. The next section shows the assume facility in action in a sample session.

4.3 An Example

We demonstrate here on a small scale how information (i.e. properties) of objects are interactively exchanged between the user and the system. We set off with two unassigned variables, a, b, and a third variable, c, assigned to the product of a and b. assume(a<0); causes first the relation a<0 to be transformed into a pair of object and property, namely a and RealRange(-infinity,Open(0)). The system then updates the ObjectTable as a side effect and returns NULL. Similarly for assume(b,OrProp(negative,RealRange(-infinity,-10.3))); only that here, the data structure OrProp first evaluates OrProp(negative,RealRange(-infinity,-10.3)) to RealRange(-infinity,Open(0)). At this stage, there are entries for a and b in the ObjectTable, but none for c. This kind of information can be retrieved with the function isgiven. isgiven builds upon this information. For instance, while isgiven(c,positive); returns false, is(c,positive); evaluates to true. This last step contains the computation of the multiplication of the two properties RealRange(-infinity,Open(0)) and RealRange(-infinity,Open(0)) with each other (yielding RealRange(Open(0),infinity)) and the validation of the inclusion of that image in positive. If we are not quite sure, whether our interpretation of positive coincides with that of the system, we just ask it, about(positive).

```
    |\^/|     MAPLE V
._|\|   |/|_. Copyright (c) 1981-1990 by the University of Waterloo.
 \  MAPLE  /  All rights reserved. MAPLE is a registered trademark of
 <____ ____>  Waterloo Maple Software.
      |       Type ? for help.

Maple session > read assumeFacility;
Maple session > c := a*b;
                                   c := a b
Maple session > assume( a<0 );
Maple session > assume( b, OrProp(negative,RealRange(-infinity,-10.3)) );
Maple session > ObjectTable[a]; ObjectTable[b];

                         RealRange(- infinity, Open(0))

                         RealRange(- infinity, Open(0))

Maple session > ObjectTable[c];

                               ObjectTable[a* b*]

Maple session > isgiven(a,negative);

                                     true
Maple > isgiven(b,real);

                                     true
Maple session > isgiven(c,positive);

                                     false

Maple session > is(c,positive);
                                     true

Maple session > about(a);
Originally a, renamed a*:
  is assumed to be: RealRange(-infinity,Open(0))

Maple session > about(negative);
negative:
  property aliased to RealRange(Open(0),infinity)
  a real range, in mathematical notation: (0,infinity)
  a known property having {RealRange(-infinity,infinity)} as immediate parents
    and {0, RealRange(Open(0),infinity)} as immediate children.

Maple session > quit;
```

An evaluation procedure for integrals supporting the assume facility makes use of this in the obvious manner. For integral expressions of the kind of our introductory example, `int(x*exp(-a*x),x=0..infinity)`, it first calls `is(a,real);` to decide whether to return unevaluated or not; in the latter case, it also calls `is(a,positive);` to decide whether to return $1/a^2$ or `infinity`.

A Implementation of the user function "is"

```
#
# FUNCTION is : Determine from the information in the ObjectTable
#               if the given property is true, false, or undecidable.
#
is := proc( obj:anything, prop:property )
local eqns, i, inds, neqns, sol;
option `Copyright 1992 Gaston Gonnet, Wissenschaftliches Rechnen, ETH Zurich`;
description `derive whether an object has the given property`;
#
if type([args],[relation]) then RETURN( procname(ConvertRelation(args)) )
elif nargs<>2 then ERROR(`invalid number of arguments`)
elif member(prop,Aliased) then RETURN( procname(obj,subs(Aliases,prop)) )
elif not type(obj,algebraic) then ERROR(`invalid arguments`)
elif isgiven(args) then RETURN(true)
#
elif assigned(ObjectTable[obj]) then
     PropInclusion(ObjectTable[obj],prop);
     if "<>FAIL then RETURN(") fi
elif assigned(ObjectTable[-obj]) then
     Function( `*`, -1, ObjectTable[-obj] );
     if " <> PropRange(BottomProp,TopProp) then
        PropInclusion(",prop);
        if "<>FAIL then RETURN(") fi fi
     fi;
#
# if the object is composed of strings, all of them with properties,
# it is ready for UpperBounded.
if WithProp(obj) then RETURN( UpperBounded(obj,prop) ) fi;
#
# represent obj as an expression of strings which have entries in the ObjectTable
eqns := { obj = _ToSolveFor };
do
   inds := indets( map( x -> op(1,x), eqns), string );
   neqns := eqns;
   for i in inds do if assigned(Involves[i]) then
         neqns := neqns union Involves[i] fi od;
   if neqns = eqns then break fi;
   eqns := neqns
   od;
if nops(eqns)=1 then RETURN(FAIL) fi;
sol := traperror( [solve( eqns, inds union {_ToSolveFor} )] );
if sol=lasterror or sol=[] then RETURN( FAIL ) fi;
if nops(sol) > 1 then ERROR(`very suspicious!!`) fi;
sol := subs(sol[1],_ToSolveFor);
if indets(sol,string) intersect inds <> {} then RETURN( FAIL ) fi;
RETURN( UpperBounded(sol,prop) )
#
end:
```

B Implementation of the property function "plus"

```
# FUNCTION `property/+` : Determine the sum of two properties.
`property/+` := proc( a, b )
local groups;
option remember,
    `Copyright 1992 Gaston Gonnet, Wissenschaftliches Rechnen, ETH Zurich`;
```

```
#
groups := { complex, GaussianInteger, RealRange(-infinity,infinity), imaginary,
          rational, integer};
if a=BottomProp or b=BottomProp then BottomProp
#
elif type([a,b],[numeric,numeric]) then a+b
elif type([a,b],[numeric,Open(numeric)]) then Open(a+op(b))
elif type([a,b],[Open(numeric),numeric]) then Open(op(a)+b)
elif type([a,b],[Open(numeric),Open(numeric)]) then Open(op(a)+op(b))
elif type([a,b],list({identical(-infinity),numeric,Open(numeric)}))then-infinity
elif type([a,b],list({identical(infinity),numeric,Open(numeric)})) then infinity
#
elif type([a,b],[numeric,RealRange]) then
     RealRange( procname(a,op(1,b)), procname(a,op(2,b)) )
elif type([a,b],[RealRange,numeric]) then procname(b,a)
elif type([a,b],[RealRange,RealRange]) then
     RealRange( procname(op(1,a),op(1,b)), procname(op(2,a),op(2,b)) )
#
elif a=0 and PropInclusion(b,complex) then b
elif b=0 and PropInclusion(a,complex) then a
#
elif member(a,groups) and PropInclusion(b,a) then a
elif member(b,groups) and PropInclusion(a,b) then b
#
else 'procname'(a,b) fi # not yet implemented, eg. PropRanges
#
end:
#
#################### SPECIAL CASES ####################
#
'property/+'( RealRange(-infinity,infinity), imaginary ) := complex:
'property/+'( imaginary, RealRange(-infinity,infinity) ) := complex:
#
'property/+'( rational, irrational ) := irrational:
'property/+'( irrational, rational ) := irrational:

  ⋮
```

References

[CGG+91a] B. Char, K. Geddes, G. Gaston, B. Leong, M. Monagan, and S. Watt. *Maple V Language Reference Manual.* Springer-Verlag, New York, 1991.

[CGG+91b] B. Char, K. Geddes, G. Gaston, B. Leong, M. Monagan, and S. Watt. *Maple V Library Reference Manual.* Springer-Verlag, New York, 1991.

[WG91] T. Weibel and G. Gonnet. An algebra of properties. In S.M. Watt, editor, *Proceedings of the 1991 International Symposium on Symbolic and Algebraic Computation: ISSAC '91*, pages 352–359, New York, 1991. ACM Press. held in Bonn, Germany.

REDUCE Meets CAMAL

J. P. Fitch

School of Mathematical Sciences
University of Bath
BATH, BA2 7AY, United Kingdom

Abstract. It is generally accepted that special purpose algebraic systems are more efficient than general purpose ones, but as machines get faster this does not matter. An experiment has been performed to see if using the ideas of the special purpose algebra system CAMAL(F) it is possible to make the general purpose system REDUCE perform calculations in celestial mechanics as efficiently as CAMAL did twenty years ago. To this end a prototype Fourier module is created for REDUCE, and it is tested on some small and medium-sized problems taken from the CAMAL test suite. The largest calculation is the determination of the Lunar Disturbing Function to the sixth order. An assessment is made as to the progress, or lack of it, which computer algebra has made, and how efficiently we are using modern hardware.

1 Introduction

A number of years ago there emerged the divide between general-purpose algebra systems and special purpose one. Here we investigate how far the improvements in software and more predominantly hardware have enabled the general systems to perform as well as the earlier special ones. It is similar in some respects to the Possion program for MACSYMA [8] which was written in response to a similar challenge.

The particular subject for investigation is the Fourier series manipulator which had its origins in the Cambridge University Institute for Theoretical Astronomy, and later became the F subsystem of CAMAL [3, 10]. In the late 1960s this system was used for both the Delaunay Lunar Theory [7, 2] and the Hill Lunar Theory [5], as well as other related calculations. Its particular area of application had a number of peculiar operations on which the general speed depended. These are outlined below in the section describing how CAMAL worked. There have been a number of subsequent special systems for celestial mechanics, but these tend to be restricted to the group of the originator.

The main body of the paper describes an experiment to create within the REDUCE system a sub-system for the efficient manipulation of Fourier series. This prototype program is then assessed against both the normal (general) REDUCE and the extant CAMAL results. The tests are run on a number of small problems typical of those for which CAMAL was used, and one medium-sized problem, the calculation of the Lunar Disturbing Function. The mathematical background to this problem is also presented for completeness. It is important as a problem as it is the first stage in the development of a Delaunay Lunar Theory.

The paper ends with an assessment of how close the performance of a modern REDUCE on modern equipment is to the (almost) defunct CAMAL of eighteen years ago.

2 How CAMAL Worked

The Cambridge Algebra System was initially written in assembler for the Titan computer, but later was rewritten a number of times, and matured in BCPL, a version which was ported to IBM mainframes and a number of microcomputers. In this section a brief review of the main data structures and special algorithms is presented.

2.1 CAMAL Data Structures

CAMAL is a hierarchical system, with the representation of polynomials being completely independent of the representations of the angular parts.

The angular part had to represent a polynomial coefficient, either a sine or cosine function and a linear sum of angles. In the problems for which CAMAL was designed there are 6 angles only, and so the design restricted the number, initially to six on the 24 bit-halfword TITAN, and later to eight angles on the 32-bit IBM 370, each with fixed names (usually u through z). All that is needed is to remember the coefficients of the linear sum. As typical problems are perturbations, it was reasonable to restrict the coefficients to small integers, as could be represented in a byte with a guard bit. This allowed the representation to pack everything into four words.

```
[ NextTerm, Coefficient, Angles0-3, Angles4-7 ]
```

The function was coded by a single bit in the `Coefficient` field. This gives a particularly compact representation. For example the Fourier term $\sin(u-2v+w-3x)$ would be represented as

```
[ NULL, "1"|0x1, 0x017e017d, 0x00000000 ]
```

or

```
[ NULL, "1"|0x1, 1:-2:1:-3, 0:0:0:0 ]
```

where `"1"` is a pointer to the representation of the polynomial 1. In all this representation of the term took 48 bytes. As the complexity of a term increased the store requirements to no grow much; the expression $(7/4)ae^3f^5\cos(u-2v+3w-4x+5y+6z)$ also takes 48 bytes. There is a canonicalisation operation to ensure that the leading angle is positive, and sin(0) gets removed. It should be noted that cos(0) is a valid and necessary representation.

The polynomial part was similarly represented, as a chain of terms with packed exponents for a fixed number of variables. There is no particular significance in this except that the terms were held in *increasing* total order, rather than the decreasing order which is normal in general purpose systems. This had a number of important effects on the efficiency of polynomial multiplication in the presence of a truncation to a certain order. We will return to this point later. Full details of the representation can be found in [9].

The space administration system was based on explicit return rather than garbage collection. This meant that the system was sometimes harder to write, but it did mean that much attention was focussed on efficient reuse of space. It was possible for the user to assist in this by marking when an expression was needed no longer, and the compiler then arranged to recycle the space as part of the actual operation. This degree of control was another assistance in running of large problems on relatively small machines.

2.2 Automatic Linearisation

In order to maintain Fourier series in a canonical form it is necessary to apply the transformations for linearising products of sine and cosines. These will be familiar to readers of the REDUCE test program as

$$\cos\theta\cos\phi \Rightarrow (\cos(\theta+\phi)+\cos(\theta-\phi))/2, \tag{1}$$
$$\cos\theta\sin\phi \Rightarrow (\sin(\theta+\phi)-\sin(\theta-\phi))/2, \tag{2}$$
$$\sin\theta\sin\phi \Rightarrow (\cos(\theta-\phi)-\cos(\theta+\phi))/2, \tag{3}$$
$$\cos^2\theta \Rightarrow (1+\cos(2\theta))/2, \tag{4}$$
$$\sin^2\theta \Rightarrow (1-\cos(2\theta))/2. \tag{5}$$

In CAMAL these transformations are coded directly into the multiplication routines, and no action is necessary on the part of the user to invoke them. Of course they cannot be turned off either.

2.3 Differentiation and Integration

The differentiation of a Fourier series with respect to an angle is particularly simple. The integration of a Fourier series is a little more interesting. The terms like $\cos(nu+\ldots)$ are easily integrated with respect to u, but the treatment of terms independent of the angle would normally introduce a secular term. By convention in Fourier series these secular terms are ignored, and the constant of integration is taken as just the terms independent of the angle in the integrand. This is equivalent to the substitution rules

$$\sin(n\theta) \Rightarrow -(1/n)\cos(n\theta)$$
$$\cos(n\theta) \Rightarrow (1/n)\sin(n\theta)$$

In CAMAL these operations were coded directly, and independently of the differentiation and integration of the polynomial coefficients.

2.4 Harmonic Substitution

An operation which is of great importance in Fourier operations is the *harmonic substitution*. This is the substitution of the sum of some angles and a general expression for an angle. In order to preserve the format, the mechanism uses the translations

$$\sin(\theta+A) \Rightarrow \sin(\theta)\cos(A)+\cos(\theta)\sin(A)$$
$$\cos(\theta+A) \Rightarrow \cos(\theta)\cos(A)-\sin(\theta)\sin(A)$$

and then assuming that the value A is small it can be replaced by its expansion:

$$\begin{aligned}\sin(\theta + A) &\Rightarrow \sin(\theta)\{1 - A^2/2! + A^4/4!\ldots\} + \\ &\quad\cos(\theta)\{A - A^3/3! + A^5/5!\ldots\} \\ \cos(\theta + A) &\Rightarrow \cos(\theta)\{1 - A^2/2! + A^4/4!\ldots\} - \\ &\quad\sin(\theta)\{A - A^3/3! + A^5/5!\ldots\}\end{aligned}$$

If a truncation is set for large powers of the polynomial variables then the series will terminate. In CAMAL the **HSUB** operation took five arguments; the original expression, the angle for which there is a substitution, the new angular part, the expression part (A in the above), and the number of terms required.

The actual coding of the operation was not as expressed above, but by the use of Taylor's theorem. As has been noted above the differentiation of a harmonic series is particularly easy.

2.5 Truncation of Series

The main use of Fourier series systems is in generating perturbation expansions, and this implies that the calculations are performed to some degree of the small quantities. In the original CAMAL all variables were assumed to be equally small (a restriction removed in later versions). By maintaining polynomials in increasing maximum order it is possible to truncate the multiplication of two polynomials. Assume that we are multiplying the two polynomials

$$\begin{aligned}A &= a_0 + a_1 + a_2 + \ldots \\ B &= b_0 + b_1 + b_2 + \ldots\end{aligned}$$

If we are generating the partial answer

$$a_i(b_0 + b_1 + b_2 + \ldots)$$

then if for some j the product $a_i b_j$ vanishes, then so will all products $a_i b_k$ for $k > j$. This means that the later terms need not be generated. In the product of $1 + x + x^2 + x^3 + \ldots + x^{10}$ and $1 + y + y^2 + y^3 + \ldots + y^1 0$ to a total order of 10 instead of generating 100 term products only 55 are needed. The ordering can also make the merging of the new terms into the answer easier.

3 Towards a CAMAL Module

For the purposes of this work it was necessary to reproduce as many of the ideas of CAMAL as feasible within the REDUCE framework and philosophy. It was not intended at this stage to produce a complete product, and so for simplicity a number of compromises were made with the "no restrictions" principle in REDUCE and the space and time efficiency of CAMAL. This section describes the basic design decisions.

3.1 Data Structures

In a fashion similar to CAMAL a two level data representation is used. The coefficients are the standard quotients of REDUCE, and their representation need not concern us further. The angular part is similar to that of CAMAL, but the ability to pack angle multipliers and use a single bit for the function are not readily available in Standard LISP, so instead a longer vector is used. Two versions were written. One used a balanced tree rather than a linear list for the Fourier terms, this being a feature of CAMAL which was considered but never coded. The other uses a simple linear representation for sums. The angle multipliers are held in a separate vector in order to allow for future flexibility. This leads to a representation as a vector of length 6 or 4;

```
Version1:  [ BalanceBits, Coeff, Function, Angles, LeftTree, RightTree ]
Version2:  [ Coeff, Function, Angles, Next ]
```

where the **Angles** field is a vector of length 8, for the multipliers. It was decided to forego packing as for portability we do not know how many to pack into a small integer. The tree system used is AVL, which needs 2 bits to maintain balance information, but these are coded as a complete integer field in the vector. We can expect the improvements implicit in a binary tree to be advantageous for large expressions, but the additional overhead may reduce its utility for smaller expressions.

A separate vector is kept relating the position of an angle to its print name, and on the property list of each angle the allocation of its position is kept. So long as the user declares which variables are to be treated as angles this mechanism gives flexibility which was lacking in CAMAL.

3.2 Linearisation

As in the CAMAL system the linearisation of products of sines and cosines is done not by pattern matching but by direct calculation at the heart of the product function, where the transformations (1) through (3) are made in the product of terms function. A side effect of this is that there are no simple relations which can be used from within the Fourier multiplication, and so a full addition of partial products is required. There is no need to apply linearisations elsewhere as a special case. Addition, differentiation and integration cannot generate such products, and where they can occur in substitution the natural algorithm uses the internal multiplication function anyway.

3.3 Substitution

Substitution is the main operation of Fourier series. It is useful to consider three different cases of substitutions.

1. Angle Expression for Angle:
2. Angle Expression + Fourier Expression for Angle:
3. Fourier Expression for Polynomial Variable.

The first of these is straightforward, and does not require any further comment. The second substitution requires a little more care, but is not significantly difficult to implement. The method follows the algorithm used in CAMAL, using TAYLOR series. Indeed this is the main special case for substitution.

The problem is the last case. Typically many variables used in a Fourier series program have had a WEIGHT assigned to them. This means that substitution must take account of any possible WEIGHTs for variables. The standard code in REDUCE does this in effect by translating the expression to prefix form, and recalculating the value. A Fourier series has a large number of coefficients, and so this operations are repeated rather too often. At present this is the largest problem area with the internal code, as will be seen in the discussion of the Disturbing Function calculation.

4 Integration with REDUCE

The Fourier module needs to be seen as part of REDUCE rather than as a separate language. This can be seen as having internal and external parts.

4.1 Internal Interface

The Fourier expressions need to co-exist with the normal REDUCE syntax and semantics. The prototype version does this by (ab)using the module method, based in part on the TPS code [1]. Of course Fourier series are not constant, and so are not really domain elements. However by asserting that Fourier series form a ring of constants REDUCE can arrange to direct basic operations to the Fourier code for addition, subtraction, multiplication and the like.

The main interface which needs to be provided is a simplification function for Fourier expressions. This needs to provide compilation for linear sums of angles, as well as constructing sine and cosine functions, and creating canonical forms.

4.2 User Interface

The creation of `HDIFF` and `HINT` functions for differentiation disguises this. An unsatisfactory aspect of the interface is that the tokens `SIN` and `COS` are already in use. The prototype uses the operator form

```
fourier sin(u)
```

to introduce harmonically represented sine functions. An alternative of using the tokens `F_SIN` and `F_COS` is also available.

It is necessary to declare the names of the angles, which is achieved with the declaration

```
harmonic theta, phi;
```

At present there is no protection against using a variable as both an angle and a polynomial variable. This will need to be done in a user-oriented version.

5 The Simple Experiments

The REDUCE test file contains a simple example of a Fourier calculation, determining the value of $(a_1 \cos(wt)+a_3\cos(3wt)+b_1\sin(wt)+b_3\sin(3wt))^3$. For the purposes of this system this is too trivial to do more than confirm the correct answers.

The simplest non-trivial calculation for a Fourier series manipulator is to solve Kepler's equation for the eccentric anomaly E in terms of the mean anomaly u, and the eccentricity of an orbit e, considered as a small quantity

$$E = u + e\sin E$$

The solution proceeds by repeated approximation. Clearly the initial approximation is $E_0 = u$. The nth approximation can be written as $u + A_n$, and so A_n can be calculated by

$$A_k = e\sin(u + A_{k-1})$$

This is of course precisely the case for which the HSUB operation is designed, and so in order to calculate $E_n - u$ all one requires is the code

```
bige := fourier 0;
for k:=1:n do <<
  wtlevel k;
  bige:=fourier e * hsub(fourier(sin u), u, u, bige, k);
>>;
write "Kepler Eqn solution:", bige$
```

It is possible to create a regular REDUCE program to simulate this (as is done for example in Barton and Fitch[4], page 254). Comparing these two programs indicates substantial advantages to the Fourier module, as could be expected.

Solving Kepler's Equation

Order	REDUCE	Fourier Module
5	9.16	2.48
6	17.40	4.56
7	33.48	8.06
8	62.76	13.54
9	116.06	21.84
10	212.12	34.54
11	381.78	53.94
12	692.56	82.96
13	1247.54	125.86
14	2298.08	187.20
15	4176.04	275.60
16	7504.80	398.62
17	13459.80	569.26
18	***	800.00
19	***	1116.92
20	***	1536.40

These results were with the linear representation of Fourier series. The tree representation was slightly slower. The ten-fold speed-up for the 13th order is most satisfactory.

6 A Medium-Sized Problem

Fourier series manipulators are primarily designed for large-scale calculations, but for the demonstration purposes of this project a medium problem is considered. The first stage in calculating the orbit of the Moon using the Delaunay theory (of perturbed elliptic motion for the restricted 3-body problem) is to calculate the energy of the Moon's motion about the Earth — the Hamiltonian of the system. This is the calculation we use for comparisons.

6.1 Mathematical Background

The full calculation is described in detail in [6], but a brief description is given here for completeness, and to grasp the extent of the calculation.

Referring to the figure 1 which gives the cordinate system, the basic equations are

$$S = (1-\gamma^2)\cos(f+g+h-f'-g'-h') + \gamma^2 cos(f+g-h+f'+g'+h') \quad (6)$$
$$r = a(1-e\cos E) \quad (7)$$
$$l = E - e\sin E \quad (8)$$
$$a = \frac{r\mathrm{d}E}{\mathrm{d}l} \quad (9)$$
$$\frac{r^2\mathrm{d}f}{\mathrm{d}l} = a^2(1-e^2)^{\frac{1}{2}} \quad (10)$$
$$R = m'\frac{a^2}{a'^3}\frac{a'}{r'}\left\{\left(\frac{r}{a}\right)^2\left(\frac{a'}{r'}\right)^2 P_2(S) + \left(\frac{a}{a'}\right)\left(\frac{r}{a}\right)^3\left(\frac{a'}{r'}\right)^3 P_3(S) + \ldots\right\} \quad (11)$$

There are similar equations to (7) to (10) for the quantities r', a', e', l', E' and f' which refer to the position of the Sun rather than the Moon. The problem is to calculate the expression R as an expansion in terms of the quantities e, e', γ, a/a', l, g, h, l', g' and h'. The first three quantities are small quantities of the first order, and a/a' is of second order.

The steps required are

1. Solve the Kepler equation (8)
2. Substitute into (7) to give r/a in terms of e and l.
3. Calculate a/r from (9) and f from (10)
4. Substitute for f and f' into S using (6)
5. Calculate R from S, a'/r' and r/a

The program is given in the Appendix.

6.2 Results

The Lunar Disturbing function was calculated by a direct coding of the previous sections' mathematics. The program was taken from Barton and Fitch [4] with just small changes to generalise it for any order, and to make it acceptable for Reduce3.4. The Fourier program followed the same pattern, but obviously used the `HSUB` operation as appropriate and the harmonic integration. It is very similar to the CAMAL program in [4].

The disturbing function was calculated to orders 2, 4 and 6 using Cambridge LISP on an HLH Orion 1/05 (Intergraph Clipper), with the three programs α) Reduce3.4, β) Reduce3.4 + Camal Linear Module and γ) Reduce3.4 + Camal AVL Module. The timings for CPU seconds (excluding garbage collection time) are summarised the following table:

Order of DDF	Reduce	Camal Linear	Camal Tree
2	23.68	11.22	12.9
4	429.44	213.56	260.64
6	>7500	3084.62	3445.54

If these numbers are normalised so REDUCE calculating the DDF is 100 units for each order the table becomes

Order of DDF	Reduce	Camal Linear	Camal Tree
2	100	47.38	54.48
4	100	49.73	60.69
6	100	<41.13	<45.94

From this we conclude that a doubling of speed is about correct, and although the balanced tree system is slower as the problem size increases the gap between it and the simpler linear system is narrowing.

It is disappointing that the ratio is not better, nor the absolute time less. It is worth noting in this context that Jefferys claimed that the sixth order DDF took 30s on a CDC6600 with TRIGMAN in 1970 [11], and Barton and Fitch took about 1s for the second order DDF on TITAN with CAMAL [4]. A closer look at the relative times for individual sections of the program shows that the substitution case of replacing a polynomial variable by a Fourier series is only marginally faster than the simple REDUCE program. In the DDF program this operation is only used once in a major form, substituting into the Legendre polynomials, which have been previously calculated by Rodrigues formula. This suggests that we replace this with the recurrence relationship.

Making this change actually slows down the normal REDUCE by a small amount but makes a significant change to the Fourier module; it reduces the run time for the 6th order DDF from 3084.62s to 2002.02s. This gives some indication of the problems with benchmarks. What is clear is that the current implementation of substitution of a Fourier series for a polynomial variable is inadequate.

7 Conclusion

The Fourier module is far from complete. The operations necessary for the solution of Duffing's and Hill's equations are not yet written, although they should not cause much problem. The main deficiency is the treatment of series truncation; at present it relies on the REDUCE WTLEVEL mechanism, and this seems too coarse for efficient truncation. It would be possible to re-write the polynomial manipulator as well, while retaining the REDUCE syntax, but that seems rather more than one would hope.

The real failure so far is the large time lag between the REDUCE-based system on a modern workstation against a mainframe of 25 years ago running a special system. The CAMAL Disturbing function program could calculate the tenth order with a maximum of 32K words (about 192Kbytes) whereas this system failed to calculate the eighth order in 4Mbytes (taking 2000s before failing). I have in my archives the output from the standard CAMAL test suite, which includes a sixth order DDF on an IBM 370/165 run on 2 June 1978, taking 22.50s and using a maximum of 15459 words of memory for heap — or about 62Kbytes. A rough estimate is that the Orion 1/05 is comparable in speed to the 360/165, but with more real memory and virtual memory.

However, a simple Fourier manipulator has been created for REDUCE which performs between twice and three times the speed of REDUCE using pattern matching. It has been shown that this system is capable of performing the calculations of celestial mechanics, but it still seriously lags behind the efficiency of the specialist systems of twenty years before. It is perhaps fortunate that it was not been possible to compare it with a modern specialist system.

There is still work to do to provide a convenient user interface, but it is intended to develop the system in this direction. It would be pleasant to have again a system of the efficiency of CAMAL(F).

I would like to thank Codemist Ltd for the provision of computing resources for this project, and David Barton who taught be so much about Fourier series and celestial mechanics. Thank are also due to the National Health Service, without whom this work and paper could not have been produced.

Appendix: The DDF Function

```
array p(n/2+2);
harmonic u,v,w,x,y,z;
weight e=1, b=1, d=1, a=1;

%% Generate Legendre Polynomials to sufficient order
for i:=2:n/2+2 do <<
  p(i):=(h*h-1)^i;
  for j:=1:i do p(i):=df(p(i),h)/(2j)
>>;

%%%%%%%%%%%%%%%% Step1: Solve Kepler equation
```

```
bige := fourier 0;
for k:=1:n do <<
  wtlevel k;
  bige:=fourier e * hsub(fourier(sin u), u, u, bige, k);
>>;

%% Ensure we do not calculate things of too high an order
wtlevel n;

%%%%%%%%%%%%%%%% Step 2: Calculate r/a in terms of e and l
dd:=-e*e; hh:=3/2; j:=1; cc := 1;
for i:=1:n/2 do <<
  j:=i*j; hh:=hh-1; cc:=cc+hh*(dd^i)/j
>>;
bb:=hsub(fourier(1-e*cos u), u, u, bige, n);
aa:=fourier 1+hdiff(bige,u); ff:=hint(aa*aa*fourier cc,u);

%%%%%%%%%%%%%%%% Step 3: a/r and f
uu := hsub(bb,u,v); uu:=hsub(uu,e,b);
vv := hsub(aa,u,v); vv:=hsub(vv,e,b);
ww := hsub(ff,u,v); ww:=hsub(ww,e,b);

%%%%%%%%%%%%%%%% Step 4: Substitute f and f' into S
yy:=ff-ww; zz:=ff+ww;
xx:=hsub(fourier((1-d*d)*cos(u)),u,u-v+w-x-y+z,yy,n)+
    hsub(fourier(d*d*cos(v)),v,u+v+w+x+y-z,zz,n);

%%%%%%%%%%%%%%%% Step 5: Calculate R
zz:=bb*vv; yy:=zz*zz*vv;

on fourier;
for i := 2:n/2+2 do <<
  wtlevel n+4-2i; p(i) := hsub(p(i), h, xx) >>;

wtlevel n;
for i:=n/2+2 step -1 until 3 do
    p(n/2+2):=fourier(a*a)*zz*p(n/2+2)+p(i-1);
yy*p(n/2+2);
```

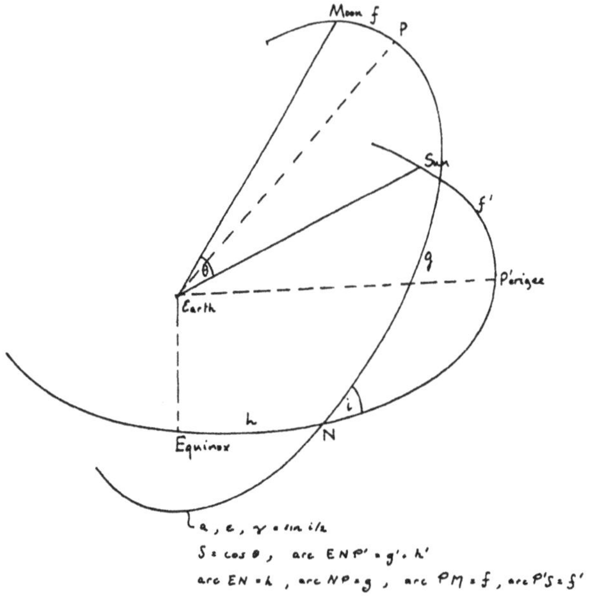

Fig. 1. Coordinate System for Earth-Moon-Sun

References

1. A. Barnes and J. A. Padget. Univariate power series expansions in Reduce. In S. Watanabe and M. Nagata, editors, *Proceedings of ISSAC'90*, pages 82–7. ACM, Addison-Wesley, 1990.
2. D. Barton. *Astronomical Journal*, 72:1281–7, 1967.
3. D. Barton. A scheme for manipulative algebra on a computer. *Computer Journal*, 9:340–4, 1967.
4. D. Barton and J. P. Fitch. The application of symbolic algebra system to physics. *Reports on Progress in Physics*, 35:235–314, 1972.
5. Stephen R. Bourne. Literal expressions for the co-ordinates of the moon. I. the first degree terms. *Celestial Mechanics*, 6:167–186, 1972.
6. E. W. Brown. *An Introductory Treatise on the Lunar Theory.* Cambridge University Press, 1896.
7. C. Delaunay. *Théorie du Mouvement de la Lune.* (Extraits des Mém. Acad. Sci.). Mallet-Bachelier, Paris, 1860.
8. Richard J. Fateman. On the multiplication of poisson series, Draft 1973.
9. J. P. Fitch. Syllabus for algebraic manipulation lectures in cambridge. *SIGSAM Bulletin*, 32:15, 1975.
10. J. P. Fitch. *CAMAL User's Manual.* University of Cambridge Computer Laboratory, 2nd edition, 1983.
11. W. H. Jeffereys. *Celestial Mechanics*, 2:474–80, 1970.

Combinatory Models and Symbolic Computation

Karl Aberer

International Computer Science Institute, 1947 Center Street,
Berkeley, CA 94704, USA.
e-mail: aberer@icsi.berkeley.edu.

Abstract. We introduce an algebraic model of computation which is especially useful for the description of computations in analysis. On one level the model allows the representation of algebraic computation and on an other level approximate computation is represented. Furthermore programs are themselves algebraic expressions. Therefore it is possible to algebraically manipulate programs of symbolic and numerical computation, thus providing symbolic computation with a firm semantic foundation and giving a natural model for mixed symbolic-numerical computation. We illustrate these facts with examples.

1 Operations of Symbolic Computation

We distinguish the following three types of operations used in computer algebra. [1]

1. Algebraic operations: `D[f,x]`, `f/g`.
2. Computational operations: `If[c>0,f,g]`, `While[a>0, .... ]`. Also composition of programs is considered as a computational operation.
3. Representation-related operations: `Expand[f]`, `Series[f,{x,x0,n}]`.

We consider programs of symbolic computations as algebraic expressions composed from these operations. The following examples show how this viewpoint can lead to strange answers of actual systems, if done "carelessly", i.e., if no account is taken of underlying structural assumptions or features of the implementation.

Example 1. Assume we want to define in the following examples specific real functions.

1. In this example we pick out a situation, that well can occur during a computation.

```
In[19]:= f[x_]:=x/0

In[20]:= f[1]
                                            1
Power::infy: Infinite expression - encountered.
                                            0
```

[1] We use in this paper *Mathematica* as one representative of a typical classical computer algebra system. All notations for computer algebra programs are hence taken out of it. Other system, that would have served as well, are *Maple*, *Macsyma* or *Reduce*. *Axiom* (former *Scratchpad*) would have made some of the examples given more difficult to present due to its type-checking mechanism.

```
Out[20]= ComplexInfinity

In[21]:= 0 * ComplexInfinity

Infinity::indt: Indeterminate expression 0 ComplexInfinity encountered.

Out[21]= Indeterminate
```

2. Here we make a reasonable attempt to define the absolute value function.

```
In[21]:= f[x_]:=If[x>0,x,-x]

In[22]:= D[f[x],x]

                (0,0,1)                      (0,1,0)
Out[22]= -If           [x > 0, x, -x] + If          [x > 0, x, -x] +

                (1,0)          (1,0,0)
>     Greater      [x, 0] If          [x > 0, x, -x]
```

3. Now we make a less reasonable attempt to define a function recursively.

```
In[23]:= f[x_]:=f[x]/2

In[24]:= f[1]

General::recursion: Recursion depth of 256 exceeded.

Out[24]=
                                               1
                                        Hold[-] Hold[f[1]]
                                               2
> ---------------------------------------------------------------------------
  9046256971665327767466483203803742801036717552003169065582623750618213253 12
```

The following is a more reasonable, but similarly successless, approach.

```
In[9]:= g[x_,n_]:=1/2 g[x,n-1]

In[10]:= g[x,0]:=x

In[11]:= g[x]:=Limit[g[x,n],n->Infinity]

In[12]:= g[1]

Out[12]= g[1]

In[15]:= D[g[x],x]

General::recursion: Recursion depth of 256 exceeded.
```

```
General::stop: Further output of General::recursion
     will be suppressed during this calculation.
^C
```

(The calculation was interrupted.)

Discussion of the examples: ∎

1. The answer is almost satisfying. Open is the question about a mathematical interpretation of `ComplexInfinity` and about the validity of the system's assumption that the result is over a complex number field.
2. The systems response to a very natural attempt to define the absolute value function is completely inappropriate. In this case a mathematical model of how to deal with piecewise defined functions is simply missing.
3. The third example illustrates that symbolic computation systems lack an algebraic mechanism to deal with functions defined as limits of infinite recursions, despite the fact, that these functions play an important role in computations in analysis.

We have to remark that these phenomena are not particular for *mathematica* but similar behavior occurs in many other systems.

The failures presented in the above examples are mainly due to the fact, that the programming constructs cannot by safely composed to new expressions, or that basic programming constructs like recursion are not provided. This again is due to the lack of an algebraic structure which contains such constructs as elements, and provides them with a consistent semantics.

2 Representation of Approximations

Now let us turn to approximative methods. We have to consider two aspects. First the results of an approximative method are often computed as limits of recursive sequences of approximations. Typical examples are Newton's method for finding zeros of a function or power-series methods for solving differential equations, where especially the latter also play an important role in symbolic computation. Therefore recursion should be provided as an algebraic construct itself. Second approximative methods make, as the name says, use of approximations. To be able to deal with approximations in an algebraic environment we need an algebraic approach to the notion of approximation and algebraic properties of approximations.

We will use the notion of *approximating an object* in the sense of having *partial knowledge of all properties of the object.*

A typical example of such partial knowledge arising from computer algebra, is found in the representation of a real algebraic number. Two kinds of partial knowledge are combined to represent a real algebraic number exactly. Let the symbol @ denote the real algebraic number. Then first a polynomial p is given of which @ is a zero: $p(@) = 0$. Second an interval $[a, b]$ is given which isolates @: $a \leq @ \leq b$. So the set of formulas $X = \{p(@) = 0, a \leq @ \leq b\}$ describes the real algebraic number completely.

We now generalize this way of representing an object (approximately or exactly). Assume an analytic structure, e.g., a totally ordered field or differential field, is defined by means of a first order theory T using a language L of first order predicate calculus with equality, not containing the special symbol @. The language L contains all the constant, operation and predicate symbols necessary for the description of the specific analytic structure. For a totally ordered field it would at least include the constant symbols $0, 1$, the operation symbols $+, -, *, ^{-1}$ and the predicate symbol $<$. We denote the set of variable-free, quantifier-free formulas in @ by $A_@$. Then we represent (partial) knowledge of an object @ of this structure by a finite or infinite set formulas X, which are satisfied by this object.

$$X = \{\phi_1(@), \phi_2(@), \ldots\} \subseteq A_@.$$

The reason to choose quantifier-free formulas is, intuitively spoken, that they are easy to verify. In the computational model we develop we will compute with such formula-sets instead of the elements of a structure. This will allow to model in an uniform way computations with approximations, by computing with formulas-sets that do not determine an object uniquely, and objects that are known exactly, e.g., by an algebraic expression τ. In this case the object is described completely by the formulas-set $\{@ = \tau\}$.

Note that this representation of knowledge gives naturally rise to a lattice structure on partial knowledge, which is simply given by the lattice induced by subset-inclusion. We will say that a formula-set X *approximates* another formula-set Y if

$$X \subseteq Y.$$

This lattice is complete since there exists a minimum, namely the empty set, and a maximum, namely $A_@$.

We have introduced three types of operations: algebraic, computational and representation-related. For each of these we will give the corresponding operations on formula-sets.

3 Algebraic Operations

One can think of an algebraic operation, like `f+g`, as an operation, that is defined by an algebraic expression containing free variables. In the example of addition the algebraic expression would be $\tau(x, y) = x + y$. Now let us have a look back on the representation of real algebraic numbers. For the addition of two real algebraic numbers we have to be able to add the corresponding intervals. Intuitively it is clear how to add two intervals.

$$\{a \leq @ \leq b\} + \{c \leq @ \leq d\} = \{a + c \leq @ \leq b + d\}.$$

This can be described more formally as an operation on formula-sets as follows. If T is the theory of totally ordered fields then we can prove

$$a \leq x \leq b, c \leq y \leq d \vdash^T a + c \leq x + y \leq b + d.$$

where $X \vdash^T \phi$ means, that the formula ϕ is in first order logic provable from the set of formulas X, using theorems from T. Hence an element, being the sum of two elements satisfying $\{a \leq @ \leq b\}$ and $\{c \leq @ \leq d\}$, has to satisfy $\{a+c \leq @ \leq b+d\}$.

This leads to a definition generalizing an operation given by a term $\tau(x_1, \ldots, x_k)$, $k \geq 0$, to an operation $\mathcal{T}^{\tau(x_1,\ldots,x_k)}$ on formula-sets $X_1, \ldots, X_k \in A_@$. $X_i|_@^{x_i}$ denotes the set of formulas where in X_i every appearance of @ is substituted by x_i.

$$\mathcal{T}^{\tau(x_1,\ldots,x_k)}(X_1, \ldots, X_k) := \{\phi(@) : X_1|_@^{x_1}, \ldots, X_k|_@^{x_k} \vdash^T \phi(\tau(x_1, \ldots, x_k))\}.$$

To simplify notation we will maintain the usual mathematical notations also for combinatory operations, as far as no confusion is possible, e.g., for $\mathcal{T}^{\tau(x,y)}(X, Y)$ we write simply $X + Y$ or for $\mathcal{T}^0$ we write 0.

The logical closure $\mathcal{Cn}(X)$ of a formula-set X is the formula-set, that contains all formulas which are logical consequences of X under $\vdash^T$. Any object satisfying all formulas in X also satisfies all formulas in $\mathcal{Cn}(X)$. So it makes no sense to distinguish between formula-sets with the same logical closure. We call such logically closed formula-sets *combinators* and we assume that a formula-set from now on always denotes its logical closure. We denote the set of combinators with $\mathcal{E}_{A_@}$. Note that an operation $\mathcal{T}^{\tau(x_1,\ldots,x_k)}$ always maps combinators into combinators. Hence we call such an operation *combinatory operation*.

The set of logically closed formula-set again forms a complete lattice, where the maximum is, as for formula-sets, $\boldsymbol{F} = A_@$, while the minimum of this lattice is now $\boldsymbol{E} = \mathcal{Cn}(\emptyset)$. We denote inclusion of combinators in this lattice by

$$X \sqsubseteq Y.$$

The lattice operations of finding the infimum and supremum of two logically closed formula-sets X, Y are then given by

$$X \sqcap Y = X \cap Y, X \sqcup Y = \mathcal{Cn}(X \cup Y).$$

4 Computational Operations

The simplest computational operation, namely *composition*, is realized by composing combinatory expressions. Let, e.g., $\mathcal{T}_1(X), \mathcal{T}_2(X)$ be combinatory operations, then their composition is given by $\mathcal{T}_1(\mathcal{T}_2(X))$.

A decision function, a so-called *conditional*, can be defined as follows:

$$\mathcal{C}^{\phi(x)}(X_1, X_2, X_3) = \begin{cases} X_2, \text{ if } \phi(@) \in X_1, \ X_1 \neq \boldsymbol{F} \\ X_3, \text{ if } \neg\phi(@) \in X_1, \ X_1 \neq \boldsymbol{F} \\ X_2 \sqcap X_3, \text{ otherwise, if } \ X_1 \neq \boldsymbol{F} \\ X_2 \sqcup X_3, \text{ otherwise,} \end{cases}$$

where $\phi(@)$ is a formula in $A_@$ and represents the condition on which the decision is based. The two first cases are easy to understand. In the third case there is not enough information to make the decision, so both alternatives may be returned.

Hence the minimal knowledge which holds for the both alternatives, namely $X_2 \sqcap X_3$, is returned. In the last case, where a contradiction occurs, both alternatives have to be returned at the same time, so one choses the best possible answer $X_2 \sqcup X_3$. This way to define decisions allows a natural exception handling by catching up computations that have failed by providing too little or too much knowledge.

We now come to the task of representing *infinite recursion*. This first requires to introduce a limit notion for infinite sequences of combinators.

Assume that a sequence of combinators X_n, $n \in \mathbb{N}$, is given. We define the limit of such a sequence as the union

$$\bigsqcup_{n \in \mathbb{N}} X_n,$$

i.e., as the union of the knowledge contained in all X_n.

Example 2. Let X_n be the nth Taylor approximation of $\exp(\iota)$ which we express as combinator by

$$X_n = 1 + \iota + \ldots + \frac{\iota^n}{n!} + \boldsymbol{D}_n,$$

where ι denotes the identity function, i.e., $\iota' = 1$. The combinator $\boldsymbol{D}_n = \{@(0) = 0, @'(0) = 0, \ldots, @^{(n)}(0) = 0\}$ allows to substitute the notation $o(\iota^n)$, which is usually used to denote the higher-order terms of a truncated power-series. Now $\frac{\iota^{n+1}}{(n+1)!}$ is approximated by $\boldsymbol{D}_n = \{@(0) = 0, @'(0) = 0, \ldots, @^{(n)}(0) = 0\}$. Furthermore $\boldsymbol{D}_n \sqsubseteq \boldsymbol{D}_{n+1}$. Hence

$$1 + \iota + \ldots + \frac{\iota^n}{n!} + \boldsymbol{D}_n \sqsubseteq 1 + \iota + \ldots + \frac{\iota^n}{n!} + \frac{\iota^{n+1}}{(n+1)!} + \boldsymbol{D}_{n+1}.$$

This shows that $X_n \sqsubseteq X_{n+1}$ or in other words that X_n is a monotonically increasing sequence of combinators. ∎

We consider in the following only limits of monotonically increasing sequences of combinators, so called *chains*. This is no essential restriction even when considering nonmonotonic number or function sequences as the following example shows.

Example 3. Assume the nonmonotonic sequence $a_i, i \geq 0$, of real numbers is given and converges to a limit a such that $a_i < a$ for i odd and $a_i > a$ for i even. Then we can easily construct out of this sequence a monotonic sequence of combinators, such that the limit of combinators describes the same point a, by

$$\boldsymbol{A}_i = \{b \leq @ \leq c : b = \min_{j \leq i} a_j, c = \max_{j \leq i} a_j\}^2.$$

∎

We define a combinator, which we want to be the limit of an infinite recursion, as the union of a recursively computed, monotone increasing sequence of combinators

[2] We assume that $\min(\emptyset) = -\infty$ and $\max(\emptyset) = \infty$. An inequality of the form $@ < \infty$ is then equivalent to a trivial formula, e.g. $@ < @ + 1$.

X_n. Let combinatory operations $T, T^1, \ldots, T^k$ and starting values $X_0, X_0^1, \ldots, X_0^k$ be given. Then the recursion for computing the X_n is defined as follows.

$$\begin{aligned} X_{n+1} &:= T(X_n, X_n^1, \ldots, X_n^k) \sqcup X_n \\ X_{n+1}^1 &:= T^1(X_n^1, \ldots, X_n^k) \\ &\vdots \\ X_{n+1}^k &:= T^k(X_n^1, \ldots, X_n^k). \end{aligned}$$

The $X_n^1, \ldots, X_n^k$ can be considered as auxiliary sequences, which have not necessarily to be monotonic, while the monotonicity of X_n is ensured by the inclusion $X_n \sqsubseteq X_{n+1}$, which holds for the main recursion. We denote the combinator $\bigsqcup_{n \in \mathbb{N}}$ computed by the recursion by

$$M^{T, T^1, \ldots, T^k}(X_0, X_0^1, \ldots, X_0^k).$$

So $M^{T, T^1, \ldots, T^k}$ is a new combinatory operation in the arguments $X_0, X_0^1, \ldots, X_0^k$. [3] The $T, T^1, \ldots, T^k$ can be considered as parameters and to obtain better readability we will denote this operation, using arguments $X, X^1, \ldots, X^k$, alternatively by

$$M(T, T^1, \ldots, T^k; X, X^1, \ldots, X^k).$$

5 Combinatory Models

We are now ready to introduce the notion of *combinatory model.* The combinatory model of a theory T is the algebraic structure

$$\mathbf{E}_{A_\bullet} =< \mathcal{E}_{A_\bullet}; T^{\tau(x_1, \ldots, x_k)}, C^{\phi(x)}, M^{T, T^1, \ldots, T^k} > .$$

Depending on the underlying theory T we can define different combinatory models. If T is the field theory then we have a *combinatory field,* if T is the differential fields theory then we have a *combinatory differential field.* In most cases it is useful to include in the equational theories of fields and differential fields ordering predicates, hence considering, e.g., totally ordered fields or partial ordered function fields. This allows the use of approximations which are described, e.g., as intervals.

We consider programs in analytic structures as algebraic expressions of combinatory models and program execution is performed by evaluating these expressions.

The following theorem is of central importance. It relates combinatory models to graph models and justifies that combinatory models are indeed algebraic structures.

Main Theorem: [Aberer, 1991B]
The combinatory model

$$\mathbf{E}_{A_\bullet} =< \mathcal{E}_{A_\bullet}; T^{\tau(x_1, \ldots, x_k)}, C^{\phi(x)}, M^{T, T^1, \ldots, T^k} >$$

is an inner algebra of the graph model $\mathbf{D}_{A_\bullet}$.

[3] In *Axiom* a similar construct is actually used for the representation of power-series.

This theorem allows us to use methods of the theory of graph models to prove properties in combinatory models. This leads to the following basic properties of combinatory operations. Proofs for these can be found in [Aberer, 1991B].

1. Continuity: Continuity is the most basic property of operations that transform information. We first define continuity for the unary case. Let X_n be a monotone increasing chain of combinators and let $\boldsymbol{T}$ be an unary combinatory operation. Then

$$\boldsymbol{T}(\bigsqcup_{n\in\mathbb{N}} X_n) = \bigsqcup_{n\in\mathbb{N}} \boldsymbol{T}(X_n).$$

In the general case, let $X_n^1, \ldots, X_n^k$ be monotone increasing chains, and let $\boldsymbol{T}$ be a k-ary combinatory operation. Then

$$\boldsymbol{T}(\bigsqcup_{n\in\mathbb{N}} X_n^1, \ldots, \bigsqcup_{n\in\mathbb{N}} X_n^k) = \bigsqcup_{n\in\mathbb{N}} \boldsymbol{T}(X_n^1, \ldots, X_n^k).$$

A simple consequence of continuity is *monotonicity.*

$$X_1 \sqsubseteq X_2 \rightarrow \boldsymbol{T}(X_1) \sqsubseteq \boldsymbol{T}(X_2).$$

2. Fixpoint properties of recursion: The main reason for restricting the definition of recursion to the case where the main recursion sequence is monotonically increasing, is to be able to prove, using continuity, algebraic relations for recursion combinators, namely fixpoint properties. In the unary case we have

$$\boldsymbol{M}(\boldsymbol{T}; X) = \boldsymbol{T}(\boldsymbol{M}(\boldsymbol{T}; X)).$$

This property can be generalized as follows. We use the shorthand notation

$$\boldsymbol{M} = \boldsymbol{M}(\boldsymbol{T}, \boldsymbol{T}^1, \ldots, \boldsymbol{T}^k; X, X^1, \ldots, X^k).$$

Let $\boldsymbol{G}$ be an m-ary combinatory operation. If $X_{n+1} = \boldsymbol{G}(X_n, \ldots, X_{n-m})$ for $n \geq m$ then

$$\boldsymbol{M} = \boldsymbol{G}(\boldsymbol{M}, \ldots, \boldsymbol{M}).$$

3. Embedding theorem: The term structure as given by the underlying theory is isomorphically represented in the term structure of combinatory operations. This is due to the following property. Let $\tau_1, \ldots, \tau_k$ be variable-free terms. Then

$$\boldsymbol{T}^{\tau(\tau_1,\ldots,\tau_k)} = \boldsymbol{T}^{\tau(x_1,\ldots,x_k)}(\boldsymbol{T}^{\tau_1}, \ldots, \boldsymbol{T}^{\tau_k}).$$

When building up a variable-free term, one applies operations to simpler variable-free terms. Informally this theorem says is that one can interchange embedding and composition.

4. Soundness: Assume that two operations $\tau_1, \tau_2 \in Te(x_1, \ldots, x_k)$ are given. Then

$$\tau_1(x_1, \ldots, x_k) = \tau_2(x_1, \ldots, x_k) \rightarrow \boldsymbol{T}^{\tau_1(x_1,\ldots,x_k)} = \boldsymbol{T}^{\tau_2(x_1,\ldots,x_k)}.$$

5. Completeness: The converse of soundness is not a general property of combinatory operations but it can be proved for certain term classes. An example of such a class are terms of the theory of differential fields, which will be introduced later, containing free variables and built up by using the constants 0, 1 and the operations $+, -, *, ^{-1}, '$. Some completeness results can be found in [Aberer, 1991A].

6. Lifting and Weakening: Certain algebraic relationships involving terms with free variables can be lifted into the combinatory model.

Let $\tau(x), \tau_i(x) \in Te(x), i = 1, \dots, k$, and $\sigma(x_1, \dots, x_k) \in Te(x_1, \dots, x_k)$ be given and $X_1, \dots, X_k \in \mathcal{E}_{A_\bullet}$. Then

$$\boldsymbol{T}^{\tau(x)}(\boldsymbol{T}^{\sigma(x_1,\dots,x_k)}(X_1,\dots,X_k)) = \boldsymbol{T}^{\tau(\sigma(x_1,\dots,x_k))}(X_1,\dots,X_k),$$
$$\boldsymbol{T}^{\sigma(x_1,\dots,x_k)}(\boldsymbol{T}^{\tau_1(x)}(X_1),\dots,\boldsymbol{T}^{\tau_k(x)}(X_k)) = \boldsymbol{T}^{\sigma(\tau_1(x_1),\dots,\tau_k(x_k))}(X_1,\dots,X_k).$$

In general such liftings lead to a loss of information, so called *weakenings*. Let

$$\tau_1(x_1,\dots,x_k) = \tau_2(x_1,\dots,x_1,\dots,x_k,\dots,x_k),$$

and $X_1, \dots, X_k \in \mathcal{E}_{A_{\bullet_1}}$. Then

$$\boldsymbol{T}^{\tau_2(x_1,\dots,x_1,\dots,x_k,\dots,x_k)}(X_1,\dots,X_1,\dots,X_k,\dots,X_k) \sqsubseteq \boldsymbol{T}^{\tau_1(x_1,\dots,x_k)}(X_1,\dots,X_k).$$

7. Diagonalization: Recursions can be diagonalized, if the underlying theory allows the representation of natural numbers. For example all theories describing extensions of $\mathbb{Q}$ allow such a representation.

8. Conditional operations: Laws for terms composed of conditional and algebraic operations can be derived. Let $\boldsymbol{T}^{\tau(x)}$ be an unary combinatory operation. Then

$$\boldsymbol{T}^{\tau(x)}(\boldsymbol{C}^{\phi(x)}(X,Y,Z)) = \boldsymbol{C}^{\phi(x)}(X,\boldsymbol{T}^{\tau(x)}(Y),\boldsymbol{T}^{\tau(x)}(Z)).$$

6 Symbolic Programs in Combinatory Models

We give now interpretations of the programs discussed in the examples of Section 2. Division by 0 is represented by the combinatory term $\boldsymbol{T}^{\frac{1}{0}}$. Now, all that can be proved about $\frac{1}{0}$ in a field theory are trivial properties that are true for all elements. So the correct interpretation is

$$\boldsymbol{T}^{\frac{1}{0}} = \boldsymbol{E}.$$

This seems also to be the intended interpretation of `Indeterminate` in *mathematica*. If an object related to the concept of infinity is needed, this can be provided in a combinatory model by using recursion, e.g., by

$$\boldsymbol{M}(\boldsymbol{T}^{2*x}; \{-1 \le @ \le 1\}).$$

The absolute value function is given as

$$\boldsymbol{A}(X) = \boldsymbol{C}^{x>0}(X, \boldsymbol{T}^{x}(X), \boldsymbol{T}^{-x}(X)).$$

Its derivative can be computed in a combinatory model as follows

$$\boldsymbol{A}'(X) = \boldsymbol{T}^{x'}(\boldsymbol{A}(X)) = \boldsymbol{C}^{x>0}(X, \boldsymbol{T}^{x'}(X), \boldsymbol{T}^{-x'}(X)).$$

Thus, e.g., $\boldsymbol{A}(0) = 0$ and $\boldsymbol{A}'(0) = -1$. In this way we can algebraically manipulate general piecewise defined functions.

Infinite recursion can be represented now in closed form. One has only to take care of the initial value, such that the sequence of approximations is monotone increasing. [4] A recursion approximating 0 is, e.g., given by the recursion operation

$$\boldsymbol{M}(\boldsymbol{T}^{\frac{x}{2}}; \{0 \leq @ \leq 1\}).$$

7 Representation-related Operations

We have omitted this class of operations so far. Operations that change the representation of an object are very useful in symbolic computation. There exist operations that preserve the complete knowledge about an object, like `Expand[x]`, and operations that lead to a loss of information, like `Series[y,{x,x0,n}]`. The second type of representation-related operations are often used for transitions to numerical computation, e.g., using `N[x]` to get a floating point representation of a real number. One property all representation-related operations have in common is that they are *retractions*. A retraction $\boldsymbol{R}$ is an operation that satisfies

$$\boldsymbol{R}(\boldsymbol{R}(X)) = \boldsymbol{R}(X).$$

Now we show two basic possibilities to introduce such retractions in combinatory models. The first operates on the level of formula-sets by restricting the formulas allowed in the representation of the object to a subset $A^r_@$ of $A_@$.

$$\boldsymbol{R}(X) = \{\phi(@) : \phi(@) \in X \wedge \phi(@) \in A^r_@\}.$$

These operations are elements of the combinatory model [Aberer, 1991A].

Example 4. A typical example for this type of retractions would be rounding of real numbers to floating point or fixed point numbers with finite precision. Then $A^r_@$ would be the finite set of formulas

$$A^r_@ = \{@ = f_i : f_i \text{ floating (fixed) point number}, i = 1, \ldots, n\}.$$

∎

Another possibility is to define the retractions as combinatory expressions, which means we introduce the retraction in form of a program.

Example 5. We illustrate this by the computing the Taylor series as a monotone increasing sequence of Taylor polynomials. This can be done by the following recursive program $\boldsymbol{Ps}(X) := \bigsqcup_{n \in \mathbb{N}} \boldsymbol{P}_n$.

$$\begin{array}{llll}
\boldsymbol{P}_{n+1} & := y_n + \boldsymbol{D}_n & \boldsymbol{P}_0 & = \boldsymbol{E} \\
y_{n+1} & := y_n + \frac{dy_n(0)}{n!} * \iota^n & y_0 & = X \\
dy_{n+1} & := dy'_n & dy_0 & = X \\
(n+1)! & := n! * n & 0! & = 1 \\
\boldsymbol{D}_{n+1} & := \iota * \boldsymbol{D}_n & \boldsymbol{D}_0 & = \{@(0) = 0\}
\end{array}$$

---

[4] Note that $\boldsymbol{M}(\boldsymbol{T}^{\frac{x}{2}}; 1) = \boldsymbol{F}$.

To show that $\boldsymbol{P}_n$ is monotone increasing we use the following three facts. [5]

1. $\boldsymbol{D}_n \sqsubseteq \boldsymbol{D}_{n+1}$: Follows by monotinicity from $\boldsymbol{D}_0 \sqsubseteq \iota * \boldsymbol{D}_0$, which is shown by $(\iota * y)'(0) = \iota'(0) * y(0) + \iota(0) * y'(0) = 1 * y(0) + 0 * y'(0)$.
2. $\boldsymbol{D}_n \sqsubseteq \iota^{n+1}$: Follows by monotinicity from $\boldsymbol{D}_0 \sqsubseteq \iota$.
3. $\boldsymbol{D}_n = \boldsymbol{D}_n + \boldsymbol{D}_n$: It is clear that $\boldsymbol{D}_0 = \boldsymbol{D}_0 + \boldsymbol{D}_0$. Using the fact that for a combinator $\boldsymbol{T}^\tau$ representing a term τ we have $\boldsymbol{T}^\tau * (X+X) = \boldsymbol{T}^\tau * X + \boldsymbol{T}^\tau * X$, we conclude by induction on n that $\boldsymbol{D}_n + \boldsymbol{D}_n = \iota * \boldsymbol{D}_{n-1} + \iota * \boldsymbol{D}_{n-1} = \iota * (\boldsymbol{D}_{n-1} + \boldsymbol{D}_{n-1}) = \iota * \boldsymbol{D}_{n-1} = \boldsymbol{D}_n$.

This allows us to conclude

$$\boldsymbol{P}_n = y_{n-1} + \boldsymbol{D}_{n-1} = y_{n-1} + \boldsymbol{D}_{n-1} + \boldsymbol{D}_{n-1} \sqsubseteq y_{n-1} + \frac{dy_n(0)}{n!} * \iota^n + \boldsymbol{D}_n = \boldsymbol{P}_{n-1}.$$

The recursive operation also defines a retraction. This is shown by using continuity:

$$\boldsymbol{Ps}(\bigsqcup_{n \in \mathbb{N}} \boldsymbol{P}_n) = \bigsqcup_{n \in \mathbb{N}} \boldsymbol{Ps}(\boldsymbol{P}_n),$$

and remarking that $\boldsymbol{Ps}(\boldsymbol{P}_n) = \boldsymbol{P}_n$. ∎

8 Symbolic Derivation of Approximate Algorithms

We want to give an illustration how the properties of combinatory operations can be used to derive symbolically approximative algorithms. The goal is to compute approximative solutions of linear differential equations, which can be represented in closed-form by using recursion operators. Although the idea of power-series plays in the following a central role, we have to point out, that the algorithm not only gives the correct solution up to a given order, but also gives inclusions of the solution in form of function intervals. This means the algorithm computes a substantial additional amount of information, which is especially useful to obtain verified bounds on the solution. Similar techniques play a role in numerical computations, see e.g., [Weissinger, 1988] for inclusion methods for differential equations.

We consider the case of a regular, linear, homogenous differential equation over the real numbers with polynomial coefficients, which can be written in the form

$$L(y) = y^{(m)} + a_{m-1}(\iota) * y^{(m-1)} + \ldots + a_1(\iota) * y' + a_0(\iota) * y = 0.$$

It is a well known fact that the solution of such an equation is analytic in a neighborhood of any point. Furthermore the coefficients of the power-series expansion of the analytic function are computable by a polynomial recursion. We will use this to construct a combinatory solution of the differential equation, in the form of a monotone increasing and recursively computed chain of approximations of the solution.

First we give the recursion for computing the power-series expansion in a way, that is especially suited for our purposes. Let $L^e(y)$ be $L(y)$ written in expanded

[5] Note that these are also properties of $o(*\iota^n)$.

form, i.e., the derivatives $y^{(i)}$ are distributed over the polynomials $a_i(\iota)$. [6] Then substitute every monomial of the form

$$c * \iota^k * y^{(i)} \quad \text{by} \quad c * \iota^k * y^{(i)}_{n+(i-m)-k},$$

where c is a constant coefficient and $y_n, \ldots$ are new variables. The y_n will turn out to be the power-series of the solution truncated at the n-th power. This substitution gives an operator

$$\widetilde{L}(y_n, \ldots, y_{n-t}),$$

where

$$t = \max_{i=1,\ldots,m-1} deg(a_i) - (i-m).$$

We make the ansatz

$$y_{n+1} = y_n + u_n, \; u_n = b_n * \iota^n, \; b_n \text{ constant}, \tag{1}$$

for the recursion and assume that

$$\widetilde{L}(y_n, \ldots, y_{n-t}) = 0. \tag{2}$$

Note that $\widetilde{L}(y, \ldots, y) = L(y) = L^e(y)$ and

$$\widetilde{L}(y, \ldots, y) \equiv L^e(y).$$

If we substitute in $\widetilde{L}$ according to (1) and use the fact that $\widetilde{L}$ is linear y_i, the equation (2) is satisfied iff

$$\widetilde{L}(u_n, \ldots, u_{n-t}) = 0.$$

This implies that the u_n satisfy a condition of the form

$$u_n = \sum_{i=0}^{t-1} p_i(n) * \iota^i * u_{n-i-1},$$

where $p_i(n)$ are polynomials in n. This gives the desired recursive computation of y_n. The starting values $u_0, \ldots, u_{t-1}$ have to be determined according to the initial values of the differential equation.

Up to now we have only reformulated well known facts. The crucial step now is to make a combinatory ansatz. We want to recursively compute an increasing chain of approximations X_n of the form

$$X_n := y_{n-t} + V_1(n) * u_{n-1} + \ldots + V_t(n) * u_{n-t} = y_{n-t} + \sum_{i=1}^{t} V_i(n) * u_{n-i}, \tag{3}$$

where $V_i(n)$ are combinators with the property const(@) $\in V_i(n)$. We substitute this ansatz into $\widetilde{L}$ and then use the following combinatory laws which follow from

[6] This means that $L(y) = L^e(y)$ but $L(y) \not\equiv L^e(y)$, where $\equiv$ denotes syntactic equivalence. This will be of importance in the combinatory embedding later.

the properties given for internal combinators (C is a constant combinator, i.e., const(@) $\in C$):

$$(X+Y)' = X' + Y',\ (C*X)' = C*X',\ C*X + C*Y \sqsubseteq C*(X+Y).$$

Using linearity in $\widetilde{L}$ we get

$$\begin{aligned}\widetilde{L}(X_n,\ldots,X_{n-t}) = \widetilde{L}(y_{n-t},\ldots,y_{n-2t}) + \widetilde{L}(V_1(n)*u_{n-1},\ldots,V_1(n)*u_{n-t-1}) + \ldots \\ +\widetilde{L}(V_t(n)*u_{n-t},\ldots,V_t(n)*u_{n-2t}) \\ \sqsubseteq V_1(n)*\widetilde{L}(u_{n-1},\ldots,u_{n-t}) + \ldots V_t(n)*\widetilde{L}(u_{n-t-1},\ldots,u_{n-2t}) = 0.\end{aligned}$$

If we assume that the X_n are a monotone increasing chain then we may conclude the following by using continuity.

$$\bigsqcup_{n\in\mathbb{N}} \widetilde{L}(X_n,\ldots,X_{n-t}) = \widetilde{L}(\bigsqcup_{n\in\mathbb{N}} X_n,\ldots,\bigsqcup_{n\in\mathbb{N}} X_n) = L^e(\bigsqcup_{n\in\mathbb{N}} X_n) \sqsubseteq 0. \tag{4}$$

Observe how we made use of the syntactical equivalence of $\widetilde{L}(y,\ldots,y)$ and $L^e(y)$, such that no weakening occurred in this step. Hence (4) implies that, if $L^e(\bigsqcup_{n\in\mathbb{N}} X_n)$ is convergent in the sense that two elements satisfying all formulas contained in this combinator are arbitrarily close, then $\bigsqcup_{n\in\mathbb{N}} X_n$ is a unique solution.

Now we want to investigate under which circumstances the sequence X_n is a chain. This will also make clear why the ansatz for X_n was chosen exactly as in (3). To do this we make the assumption that we are only interested in the solution on a certain interval $C = \{a \le @ \le b\}$, where $-1 \le a \le b \le 1$. This will allow us to use the inclusion $C \sqsubseteq \iota^k$. Then we get

$$\begin{aligned}X_{n+1} &= y_{n-t+1} + \sum_{i=0}^{t-1} V_{i+1}(n)*u_{n-i} \\ &= y_{n-t} + u_{n-t} + \sum_{i=0}^{t-1} V_{i+1}(n)*u_{n-i} \\ &= y_{n-t} + u_{n-t} + V_1(n)*\sum_{i=1}^{t} p_i(n)*u_{n-i}*\iota^i + \sum_{i=1}^{t-1} V_{i+1}(n)*u_{n-i} \\ &= y_{n-t} + u_{n-t}*(1+V_1(n)*p_t(n)*\iota^t) + \sum_{i=1}^{t-1} u_{n-i}*(V_{i+1}(n)+V_1(n)*p_i(n)*\iota^i) \\ &\sqsupseteq y_{n-t} + u_{n-t}*(1+V_1(n)*p_t(n)*C) + \sum_{i=1}^{t-1} u_{n-i}*(V_{i+1}(n)+V_1(n)*p_i(n)*C)\end{aligned}$$

The X_n form a chain if the last line is an approximation of

$$X_n = y_{n-t} + V_1(n)*u_{n-1} + \ldots + V_t(n)*u_{n-t} = y_{n-t} + \sum_{i=1}^{t} V_i(n)*u_{n-i}.$$

This is the case if the following system of inclusions is satisfied.

$$V_t(n-1) \sqsubseteq 1 + V_1(n) * p_t(n) * C$$
$$V_i(n-1) \sqsubseteq V_{i+1}(n) + V_1(n) * p_i(n) * C, \; i = 1, \ldots, t-1.$$

These inclusions are satisfied if the corresponding equations are satisfied. The corresponding system of equations looks like a linear system of equations. In fact if we assume that C is a real number and the V_i are independent of n we have a linear system of t equations in t unknowns, which (in general) [7] is nondegenerate. We leave it as an open question whether the combinatory system is solvable in general for any linear differential equation, although this seems to be very likely following the above arguments.

We illustrate the algorithm for a concrete example which was computed by using *mathematica*. Take the differential equation

$$y''' + y' + x * y = 0, \; y(0) = 1, \; y'(0) = 1, \; y''(0) = 2.$$

The recursion is computed as

```
                 4                       2                     2
                j  u[-4 + n]          2 j  u[-2 + n]        j  n u[-2 + n]
u[n] -> -(---------------) + --------------- - ---------------
                     2    3                2    3                2    3
                2 n - 3 n  + n        2 n - 3 n  + n        2 n - 3 n  + n
```

The equation to be satisfied are

```
{V[2, n] == V[1, -1 + n],

     Int[-1, 0] V[1, n]
>    ------------------ + V[3, n] == V[2, -1 + n], V[4, n] == V[3, -1 + n],
        (-1 + n) n

          Int[-1, 0] V[1, n]
>    1 + ------------------- == V[4, -1 + n]}
          (-2 + n) (-1 + n) n
```

One solution of this system is

```
V[1,n_]:=Int[0,1];
V[2,n_]:=Int[0,1];
V[3,n_]:=Int[1-1/(n+2)/(n+1)/n,1];
V[4,n_]:=Int[1-1/(n+1)/(n-1)/n,1];
```

[7] It is easy to compute in this case the determinant and to see that only for exceptional values of C and n the system degenerates.

The first few iterates are then as follows.

```
            3       1        2                  5           23
X[3]    = J   Int[-(-), 0] + J   Int[0, 1] + Int[-, 1] + J Int[--, 1]
                    6                           6           24


               3       1        4        1             23        2    59
X[4]    = 1 + J   Int[-(-), 0] + J   Int[-(-), 0] + J Int[--, 1] + J   Int[--, 1]
                       6                 8             24               60

                   3       1     119      4       1          5      1
X[5]    = 1 + J + J   Int[-(-), -(---)] + J   Int[-(-), 0] + J   Int[-(---), 0] +
                           6     720              8                 120

                   2   3       1     119      4       1     209
X[6]    = 1 + J + J  + J   Int[-(-), -(---)] + J   Int[-(-), -(----)] +
                               6     720              8     1680

     5       1          6       1
>   J   Int[-(---), 0] + J   Int[-(---), 0]
            120                 240
                       3
                   2   J    4       1     209       5       1       67
X[7]    = 1 + J + J  - -- + J   Int[-(-), -(----)] + J   Int[-(---), -(-----)] +
                       6            8     1680              120     8064

     6       1          7       1
>   J   Int[-(---), 0] + J   Int[0, ----]
            240                 1008
```

Note that the iterates are monotone increasing combinators and how the sizes of the intervals shrink.

9 Conclusion

We have presented an algebraic approach to deal with approximate computation. This allows us to use the *qualitative* properties of approximations in an algebraic computing environment. Additionally this approach includes other important concepts for computing in analysis, like conditional functions, infinite recursion and mechanisms for exception handling. The *quantitative* properties of approximation, which lead to questions of convergence and complexity, are part of ongoing work [Aberer & Codenotti, 1992], and are important steps toward further integration of concepts from numerical and symbolic computation.

References

[Aberer, 1990] Aberer, K. (1990). Normal Forms in Function Fields. *Proceedings ISSAC '90*, 1-7.

[Aberer, 1991A] Aberer, K. (1991): Combinatory Differential Fields and Constructive Analysis. *ETH-Thesis*, **9357**.

[Aberer, 1991B] Aberer, K. (1991): Combinatory Differential Fields: An Algebraic Approach to Approximate Computation and Constructive Analysis. *TR-91-061, International Computer Science Institute, Berkeley.*.

[Aberer & Codenotti, 1992] Aberer, K., Codenotti, B. (1992). Towards a Complexity Theory of Approximation. *TR-92-012, International Computer Science Institute, Berkeley.*

[Barendregt, 1977] Barendregt, H. (1977): The type free lambda calculus. *Handbook of Mathematical Logic, ed. Jon Barwise, North Holland.*

[Blum *et al.*, 1989] Blum, L., Shub, M., Smale, S. (1989). On a Theory of Computation and Complexity over the Real Numbers: NP-Completeness. Recursive Functions and Universal Machines, *Bulletin of AMS, Vol.* **21**, *No. 1.*

[Buchberger *et al.*, 1983] Buchberger, B., Collins, B., Loos, R. (1983). Computer Algebra-Symbolic and Algebraic Computation. *Springer, Wien, New York.*

[Davenport *et al.*, 1988] Davenport, J.H., Siret, Y. and Tournier, (1988). Computer Algebra. *Academic Press, N.Y.*.

[di Primo, 1991] di Primo, B., (1991). Nichtstandard Erweiterungen von Differentialkörpern. *ETH-Thesis*, **9582**.

[Engeler, 1981A] Engeler, E., (1981). Metamathematik der Elementarmathematik. *Springer Verlag.*

[Engeler, 1981B] Engeler, E. (1981). Algebras and Combinators. *Algebra Universalis*, 389-392.

[Engeler, 1988] Engeler, E. (1988). A Combinatory Representation of Varieties and Universal Classes. *Algebra Universalis*, **24**.

[Engeler, 1990] Engeler, E. (1990). Combinatory Differential Fields. *Theoretical Computer Science* **72**, 119-131.

[Fehlmann, 1981] Fehlmann, T. (1981). Theorie und Anwendung des Graphmodells der kombinatorischen Logik. *Berichte des Instituts für Informatik der ETH* **41**.

[Kaplansky, 1957] Kaplansky, I. (1957). An Introduction to Differential Algebra. *Paris: Hermann.*

[Kaucher, 1983] Kaucher, E. (1983): Solving Function Space Problems with Guaranteed Close Bounds. *In Kulisch, U. and Miranker, W.L.: A New Approach to Scientific Computation , Academic Press, New York, p 139-164.*

[Maeder, 1986] Mäder, R. (1986). Graph Algebras, Algebraic and Denotational Semantics. *ETH-Report* **86-04**.

[Pour-El & Richards, 1989] Pour-El, E., Richards, I.J. (1989). Computability in Analysis and Physics. *Springer.*

[Risch, 1969] Risch, R. (1969). The Problem of Integration in Finite Terms. *Transactions AMS, Vol. 139, p. 167-189.*

[Stetter, 1988] Stetter, H.J., (1988). Inclusion Algorithms with Functions as Data. *Computing, Suppl.6, p.213-224.*

[Traub *et al.*, 1988] Traub, J.F., Wasilkowski G.W., Wozniakowski H., (1988). Information Based Complexity. *Academic Press, New York.*

[Weihrauch, 1980] Weihrauch, K. (1980). Rekursionstheorie und Komplexitätstheorie auf effektiven CPO-S. *Informatikberichte FernUniversität Hagen, 9/1980.*

[Weissinger, 1988] Weissinger, J. (1988). A Kind of Difference Methods for Enclosing Solutions of Ordinary Linear Boundary Value Problems. *Computing, Suppl. 6*, 23-32.

[Wolfram, 1988] Wolfram, S. (1988). Mathematica. *Addison-Wesley Publishing Company.*

A UNIFORM APPROACH TO DEDUCTION AND AUTOMATIC IMPLEMENTATION

Sergio Antoy (*), Paola Forcheri (**), Maria Teresa Molfino (**), Carlo Schenone
(*) Portland State University, Dept. of Computer Science - Portland, OR 97207, USA
(**) Istituto per la Matematica Applicata del C.N.R. - Via L.B.Alberti, 4 - Genova, Italy

ABSTRACT

Structures defined by a finite set of equations can be handled both from a mathematical and a computational point of view by using term rewriting. It is therefore worthwhile to follow this approach in realizing a programming environment which integrates the capability of defining mathematical objects, which can be directly implemented, with that of automatically verifying their properties. From the user point of view, such a system presents several advantages: first of all, actual computational entities are defined by using an abstract formalism, thus being independent of the actual programming environment; second, the use of this formalism permits proving mathematical properties of computational entities, so that the correctness of the specification and of its execution are guaranteed; finally, the executable code can be automatically derived from the formal definition of a computational entity: formal definitions become in this way tools for rapid prototyping.
On this basis, we developed MEMO (Manipulating Executable Mathematical Objects), a system which integrates the capability of automatically deriving Prolog code from algebraic specifications with that of proving their properties by using rewriting and induction. A prototype version of the system runs on a SUN 3/60 under the UNIX operating system.

1 INTRODUCTION

Powerful methods and techniques, based on different computing paradigms, such as numerical computation, symbolic computation, automatic deduction and computer graphics, are available for mathematical problem solving.

-----

* This work has been partially supported by "Progetto Finalizzato Sistemi Informatici e Calcolo Parallelo" of C.N.R. under the grant n.90.00750.69 and NSF n. CCR-8908565

However, as different computing facilities are usually available from different software tools, a considerable amount of time and effort is required to learn how to deal with them and users tend to focus on technical problems rather than on topics of interest. These difficulties can be overcome by using sofware systems which integrate different computing facilities in a single computational environment [Dewar and Richardson, 1990].

These considerations led us to design and develop MEMO, an interactive system which integrates, in the same computational environment, the capability of performing formal proofs in equational systems with that of describing the computational processes operating on them. The system provides facilities for building equational models of mathematical domains; for the generation and verification of properties of a given domain; for the analysis of the links between different domains; for the automatic implementation, in Prolog, of a given domain.
MEMO is part of the TASSO project [Miola 1990], co-ordinated by A. Miola and supported by the National Research Council of Italy under the project "Sistemi informatici e calcolo parallelo".

2 THE MEMO SYSTEM

Under suitable hypothesis, algebraic specifications of abstract data types can be given a computational interpretation, by considering the set of equations of a specification as a term rewriting system R. If R is noetherian and confluent this interpretation allows us to automatically obtain code from a specification by a process of direct implementation and, at the same time, to carry out proofs of properties of a specification by using induction and rewriting.
MEMO was designed on the pattern of this approach. The domains on which the system operates are described by means of a language based on algebraic specifications of abstract data types (see paragraph 2.1); the properties of a domain are proved using an automated theorem prover somewhat similar to a many-sorted version of the Boyer and Moore theorem prover (see paragraph 2.2); the automatic implementation is based on the idea of 'direct' implementation and it is realised through the translation process described in paragraph 2.3; the validation of a specification with respect to its computational properties avails itself of the concept of critical pairs and varies its applications depending on the domain at hand (see paragraph 2.4).

A menu-based interface allows easy access to the capabilities of the system (see paragraph 2.5). The possibilities provided by the menu enable the user to solve the following problems:

- To find examples of data domains which satisfy the defining axioms for a particular class of algebraic structures.
- Given a domain, to verify properties.
- Given a domain, to find out properties.
- To classify all structures which satisfy a particular set of defining axioms.
- Given a domain A, to identify A in terms of known classifications.
- Given a data domain and a structure domain, to analyse their relationships.
- To compare the proofs of the same properties depending on the knowledge used.
- To compute with individual elements of an algebraic domain.
- To verify if the defining equations of an algebraic domain can be directly executed, that is to verify if a program can be automatically derived from the axioms of the specification.

The MEMO system is outlined below. We assume the reader familiar with the basic notions of algebraic specification (Ehrig and Mahr 1985), term rewriting (Dershowitz and Jouannaud, 1990), computational interpretation of algebraic specifications via term rewriting (Avenhaus and Madlener, 1990).

2.1 The specification language

The MEMO system operates on data domains, such as natural numbers, integer numbers and so on, and on structure domains, such as semigroup, monoid, group and so forth. The specification (S,F,A) of a domain N consists of a signature (S,F), where S is a set of sort names (the base sets) and F is a set of function symbols with their arity, and of a set A of defining equations. Data domains are characterised by having a signature F partitioned in two disjoint subsets, say C and D, so that functions of C, called constructors, allow the attainment of the objects of N, and D is the set of the operations defined on the objects of N. A domain can be hierarchically defined, or it can be viewed as the union of previously defined domains.

Figure 1 shows an example of definition of the structure domain *group*. In this example, the phrase *USES CLASS semigroup* enables the *group* specification to inherit the sort names, operation symbols and equations of the *semigroup* specification.
The *group* specification consists of a base set *sm,* a composition operation

comp:smxsm->sm which is associative, has an identity element *e* and an inverse operation *i*. In the equations of the specification, upper case letters indicate universally quantified variables. The symbol => indicates that these equations are "oriented".
The abstract syntax of the language in a BNF-like notation is shown in Figure 2.

```
CLASS semigroup IS
 SORT sm;
 OPERATION comp: sm,sm -> sm IS AXIOM
 comp(comp(X,Y),Z) => comp(X,comp(Y,Z));
END

CLASS group IS
 USES CLASS semigroup;
OPERATION e: -> sm IS AXIOM
 comp(e,X) => X;
OPERATION i: sm -> sm IS AXIOM
 comp(i(X),X) => e;
END
```

Figure 1. Hierarchical specification of the structure domain *group*.

```
<description> ::= <object def> | <class def>
<object def> ::= OBJECT <object name> IS <name def> <operation def> END
<name def> ::= <enrichment def> | {<used objects list>;} {<sort def list>
<enrichment.def> ::= ENRICHMENT OF <object name>;
<used object list> ::= USES <objects list>
<objects list> ::= OBJECT <object name>, <objects list> | OBJECT <object name>
<sort def list> ::= <sort def> <sort def list> | <sort def>
<sort def> ::= SORT <sort name> : CONSTRUCTOR <range> {<quotient def>}
<quotient def> ::= AXIOM <axiom list>
<operation def> ::= OPERATION <range list> IS AXIOM <axiom list>
<range list> ::= <range> ; <range list> | range
<range> ::= <range name> : <sort name list> -> <sort name>
<sort name list> ::= <sort name> , <sort name list> | <sort name>
<axiom list> ::= <axiom> <axiom list> | <axiom>
<axiom> ::= <function> => <functor>
<function> ::= <function name> ( <function arguments> )
<functor> ::= <variable name> | <constant name> | <function>
<function arguments> ::= <functor>, <function arguments> | functor
<class def> :: CLASS <class name> IS <used class or sort list> ; <operations def> END
<used class or sort list> ::= <used class list> | <sort list>
<used class list> ::= USES <classes list>
<classes list> ::= CLASS <class name>, <classes list> | CLASS <class name>
<sort list> ::= SORT <sort name> <sort list> | SORT <sort name>
```

Figure 2. The syntax of the language

2.2 Proof of properties

MEMO is capable of proving properties of a domain and of proving associations between a data domain and a structure domain.
As stated before, the proof of a property is based upon a many-sorted version of the Boyer and Moore theorem prover [Antoy, 1987]. An example of a printout of a proof is given in the Appendix.
The proof of an association is carried out by verifying if this association represents a morphism between the involved domains: the functional association between the operators of the structure domain and those of the data domain is verified; if so, the equations of the structure domain are particularised to the data domain and an attempt is made to prove them.

The environment of any property we intend proving is a description (S,F,A U E) of a domain N, where A is the set of defining equations of the specification (S,F,A) of N, and E is the set of known properties of N. The set of rewriting rules associated to A U E must be proved to be canonical. E is the disjoint union of the set B of proved properties and of the set P of properties inherited through the links, if already proved, between N and another domain.

THE NUMERICAL DOMAIN INTSUM		
Defining equations (A)	**Properties (E)**	
	Proved (B)	**Inherited (P)**
add(0,X)=X		add(opp(X),add(X,Y))=Y
add(succ(X),Y)=succ(add(X,Y))	add(opp(X),X)=0	add(X,0)=X
add(pred(X),Y)=pred(add(X,Y))	add(add(X,Y),Z)=add(X,add(Y,Z))	opp(opp(X))=X
succ(pred(X))=X		add(X,opp(X))=0
pred(succ(X))=X		add(X,add(opp(X),Y))=Y
opp(0)=0		opp(add(X,Y))=add(opp(Y),opp(X))
opp(succ(X))=pred(opp(X))		
opp(pred(X))=succ(opp(X))		

Figure 3. Description of the INTSUM domain seen as a group with respect to the sum.

Figure 3 shows the set A U E associated to the data domain *INTSUM*, seen as a *group*. The *INTSUM* specification consists of a base set *int*, three constructors: *0:->int; succ:int->int; pred:int->int*, a binary operation *add:int x int->int*, and an unary operation *opp: int ->int*, whose meaning is given by the defining equations shown in the figure. B is made up of the properties proved to verify that *INTSUM* is a group when *add* is interpreted as the composition law, *0* as the identity element with respect to the composition, and *opp* as the inverse operation. The properties of P are inherited from the group structure. They are automatically obtained by particularizing the properties which correspond to the rewrite rules obtained by completion of the equations of the group structure shown in Figure 1.

2.3 Direct implementation

The MEMO system provides for the direct implementation of data domains in Prolog. The direct implementation of a specification is a process conceptually similar to that traditionally carried out by programmers. The adjective "direct" means that this implementation is based on a procedural interpretation (via term rewriting) of the algebraic equations, and it can be obtained automatically [Antoy, Forcheri, Molfino 1990; Guttag, Horowitz, Musser 1978]. To be suitable for our direct implementation, a term rewriting system must be canonical. In these systems, every term has a single normal form, that is a representation that cannot be further reduced by the defining equations of the specification. In a certain sense, the normal form of a term is its value. According to this viewpoint, we regard a term t which is not in normal form as the invocation of a computation whose result is the normal form of t. The encoding of data domains into Prolog programs in based on this idea.

Terms in normal form are translated into identical Prolog terms.

The scheme for translating defined operations of the specification language into Prolog predicates is based on the function τ which maps abstract terms into strings of Prolog terms. The translation scheme is briefly described in the following.

If f is an operation with n arguments of the specification, the direct implementation in Prolog of f is a predicate f with $n+1$ arguments. The additional argument of f is used to return the result of f applied to the other arguments. For each of the operation's defining equations, a single Horn clause is generated, and no fetching of occurrences is necessary since the pattern matching feature of Prolog is identical to that of our specification language. In order to describe the details of the translation we introduce a few notational conventions. P is the set of Prolog well-formed terms and P^+ is the set of non-null strings over P. The comma symbol is overloaded, it denotes both separation of string elements and concatenation of strings, i.e. if $x = x_1,.....,x_i$ and $y = y_1,.....,y_j$ are strings, with $i, j \geq 0$, then $x,y = x_1,....,x_i,y_1, \ldots ,y_j$. If f is a function whose range is a set of non-null strings, then $\dot{f}(x)$ is the last element of $f(x)$ and $\bar{f}(x)$ is $f(x)$ without its last element. Combining the previous two notations, we have $f(x)=\bar{f}(x),\dot{f}(x)$. Symbols of the language are mapped into Prolog symbols with the same spelling. This abuse of notation, which simplifies our formulas, is resolved by the context in which symbols occur and, whenever possible, by the font in which symbols are typed. T is a "fresh" Prolog variable, i.e. a variable which does not occur elsewhere.

$$(1)\quad \tau(t)=\begin{cases} \mathrm{X} & \text{if } t=X \text{ and } X \text{ is a variable} \\ \bar{\tau}(t_1),\ldots,\bar{\tau}(t_k),\mathrm{c}(\dot{\tau}(t_1),\ldots,\dot{\tau}(t_k)) & \text{if } t=c(t_1,\ldots,t_k) \text{ and } c \text{ is a constructor;} \\ \bar{\tau}(t_1),\ldots,\bar{\tau}(t_k),\mathrm{f}(\dot{\tau}(t_1),\ldots,\dot{\tau}(t_k),\mathrm{T}),\ \mathrm{T} & \text{if } t=f(t_1,\ldots,t_k) \text{ and } f \text{ is an operation.} \end{cases}$$

The definition (1) is extended by (2) to include the defining equations in its domain and consequently, the set of Horn clauses in its range. τ is extended as follows.

$$(2)\quad \tau(f(t_1,\ldots,t_k)\rightarrow t)=\begin{cases} \mathrm{f}(t_1,\ldots,t_k,\dot{\tau}(t)). & \text{if } \bar{\tau}(t) \text{ is null;} \\ \mathrm{f}(t_1,\ldots,t_k,\dot{\tau}(t))\text{:-}\bar{\tau}(t). & \text{otherwise.} \end{cases}$$

τ associates a Prolog predicate f to each operaration f of the language. The correctness of our methods is proved in [Antoy 1989]. More precisely, t is the value (normal form) of $f(t_1,\ldots,t_k)$ if and only if $\mathrm{f}(t_1,\ldots,t_k,t)$ is satisfied.
This approach is substantially similar to the vEM method discussed in [van Emden, Yukawa 87]. An example of the result of this process is shown in Figure 4, concerning the operation *max* which computes the maximum between two natural numbers. The constructors of the natural numbers are indicated by the symbols *0* and *s* respectively.

Defined operation	**Prolog code**
OPERATION max:nat,nat->nat IS AXIOM	
max(0,0)=>0	max(0,0,0).
max(0,s(X))=>s(X)	max(0,s(X),s(X)).
max(s(X),0)=>s(X)	max(s(X),0,s(X)).
max(s(X),s(Y))=>s(max(X,Y))	max(s(X),s(Y),s(T)):-max(X,Y,T).

Figure 4. The operation *max* and its translation into PROLOG code.

2.4 The validation process

In MEMO, different validation problems are taken into account: 1) to verify the executability of the specification of a data domain; 2) to validate the environment of a proof with respect to the computational model adopted.

1) The executability is guaranteed by checking the following conditions: operations of the specifications are completely defined, that is, for every defined function f, one and only one defining equation is applicable to any tuple of ground elements; the defining equations of the specifications can be interpreted as a canonical term rewriting system, that is, a computation terminates for every input with a unique result.
Two different mechanism are used to verify these properties. These mechanisms are based, respectively, upon the idea of "a priori" and "a posteriori" validation of a specification.
As far as the "a priori" validation is concerned, a specification oriented editor has been realised. By using this editor, the user automatically builds executable specifications. The central core of the editor comprises two procedures, called *binary choice* and *recursive reduction* respectively, which implement design strategies for specifications [Antoy, 1990].
Binary choice is used to generate the defining equations of an operation. Its application guarantees that one and only one axiom is applicable to any tuple of ground elements.
Recursive reduction is used to generate operations defined by using recursion.
As far as the "a posteriori" validation is concerned, a system has been devised to reduce the problem of verifying the executability down to checking for completeness of the definition of the operations and confluence and noetherianity of the system of rewrite rules associated with the specification [Forcheri and Molfino, 1990].

2) The process of validating the environment of a proof depends on the specification at hand.
As regards specifications which represent structure domains, we try to complete the set of rules corresponding to the equations of the specification, into a noetherian and confluent system. This process is carried out by using an incremental version of the Knuth-Bendix completion procedure. To orient the rules, the user can choose among different ordering methods [Dershowitz, Jouannaud 1990; Huet 1981].
As regards specifications which represent data domains, we build the set obtained by adding to the defining equations of the specification the properties already proved and try to interpret this set as a canonical term rewriting system.

2.5 Access to the system capabilities

The system capabilities are accessed by means of an interface, which is based on a menu.

The entries of the menu pertain to five classes:

a) entries for specifying formal domains (DIRECTORY, EDIT, MAND, NAME);

b) entries for verifying the computational properties of a domain (COMPLETE, VERIFY);

c) entries for analysing mathematical properties of a domain (PROVE, SPECIALIZE);

d) an entry for coding the specification of a domain (TRANSLATE);

e) utilities (HELP, QUIT, SHELL).

The entries and the corresponding functions are shown in Figure 5.

Entries	**Functions**
DIRECTORY	To choose among existing files.
EDIT	To specify a domain by using the language described in 2.1. Different editors can be used.
MAND	To analyse syntactic errors.
NAME	To define the working file.
COMPLETE	To complete the equational description of an algebraic structure.
VERIFY	To verify if the set of equations of a data domain is a canonical term rewriting system.
PROVE	To verify properties of a given domain.
SPECIALIZE	To analyse a given association between an abstract and a data domain.
TRANSLATE	To derive Prolog code from an executable specification of a data domain, through the process described in 2.3.
HELP	To guide the user in the use of the MEMO system.
QUIT	To exit from the system.
SHELL	To activate the commands of the operating system.

Figure 5. The entries of the MENU

3 REMARKS

In this paper, we have described the MEMO system. The design of the system emphasizes the addition of deductive capabilities to a rapid prototyping technique. As regards deduction, a particularly interesting feature is the carrying out of different proof processes for the same property, depending on the knowledge used. As regards the automatic implementation of specifications, it must be noted that the Prolog code obtained via direct implementation is inefficient. At present, our line of work consists in code optimizing.

REFERENCES

Antoy, S., (1987), "Automatically Provable Specifications", Ph.D. Thesis Department of Computer Science, University of Maryland

Antoy, S., (1989), "Algebraic Methods in Prolog Programming", Virginia Tech, TR89-5, March 1989

Antoy, S., (1990), "Design strategies for rewrite rules", Proceedings of 2th Int. Workshop on Conditional and Typed Rewriting, LNCS 516, Springer-Verlag, pp.333-341

Antoy, S., Forcheri, P., Molfino, M.T.,(1990), "Specification-based code generation", Proceedings of the Twenty-third Annual Hawai Int. Conference on Systems Sciences, IEEE Computer Society Press, pp.165-173

Antoy, S., Forcheri, P., Molfino, M.T., Zelkowitz, M.,(1990), "Rapid prototyping of Systems Enhancements", Proceedings of the First Int. Conference on Systems Integration, IEEE Computer Society Press, pp.330-336

Avenhaus, J., Madlener K., (1990), "Term Rewriting and Equational Reasoning", in Formal Techniques in Artificial Intelligence, Banerji R.B. (ed.), North Holland, pp.1-43

Boyer, R.S., Moore J.S., (1975), "Proving Theorems about LISP Functions", JACM 22-1, pp.129-144

Butler, G., Cannon, J., (1990), "The Design of Cayley - A Language for Modern Algebra", Proceedings of DISCO90, Design and Implementation of Symbolic Computation Systems, Miola A. (Ed.), LNCS 429, Springer-Verlag, pp.10-19

Dershowitz, N., Jouannaud, J.P., (1990), "Rewrite systems", in Handbook of Theoretical Computer Science, vol.B, (Van Leeuwen, J., ed.), Elsevier, pp.243-320

Dewar, M.C., Richardson, M.G., (1990), "Reconciling Symbolic and Numeric

Computation in a Practical Setting", Proceedings of DISCO90, Design and Implementation of Symbolic Computation Systems, Miola A. (Ed.), LNCS 429, Springer-Verlag, pp.195-204

Ehrig, H., Mahr, B., (1985), "Fundamentals of Algebraic Specifications 1", Springer-Verlag, Berlin,

Forcheri, P., Molfino, M.T., (1990), "Educational Software Suitable for Learning Programming", Proceedings of CATS'90, CIMNE-Pineridge Press, pp.161-164

Guttag J.V., Horowitz E., Musser D.R., (1978), "Abstract Data Types and Software Validation", ACM Communications, vol.21, N.12, pp.1048-1063

Huet, G., (1981), "A Complete Proof of Correctness of Knuth-Bendix Completion Algorithm", Journal of Computer and System Sciences, N.23, pp.11-21

Miola, A., (1990), "Tasso - A System for Mathematical Problems Solving", in Computer Systems and Applications, Balagurusamy E., Sushila, B., (Eds.), Tata McGraw-Hill

Maarten H. van Emden and Keitaro Yukawa, (1987), "Logic Programming with Equations", The Journal of Logic Programming, 4, pp.265-288.

APPENDIX

Print_out of the proof of the property

$$or(and(odd(X),odd(Y)),and(even(X),even(Y)))=even(add(X,Y))$$

in the environment *(S,F, A U E)* where:

$$S = \left\{ \begin{array}{l} nat \\ boolean \end{array} \right\}$$

$$F = \left\{ \begin{array}{l} 0{:}\rightarrow nat \\ succ:nat \rightarrow nat \\ add:natxnat \rightarrow nat \\ true:\rightarrow boolean \\ false:\rightarrow boolean \\ and:booleanxboolean \rightarrow boolean \\ or:booleanxboolean \rightarrow boolean \\ even:nat \rightarrow boolean \\ odd:nat \rightarrow boolean \end{array} \right\}$$

$$A = \left\{ \begin{array}{l} add(0,X)=X \\ add(succ(X),Y)=succ(add(X,Y)) \\ and(true,X)=X \\ and(false,X)=false \\ or(true,X)=true \\ or(false,X)=X \\ even(0)=true \\ even(succ(X))=odd(X) \\ odd(0)=false \\ odd(succ(X))=even(X) \end{array} \right\}$$

$E=\Phi$

```
The axioms are:
 natbool_1 and(true,A0) -> A0
 natbool_2 and(false,A1) -> false
 natbool_3 or(true,A2) -> true
 natbool_4 or(false,A3) -> A3
 natbool_5 add(0,A4) -> A4
 natbool_6 add(succ(A5),A6) -> succ(add(A5,A6))
 natbool_7 even(0) -> true
 natbool_8 even(succ(A7)) -> odd(A7)
 natbool_9 odd(0) -> false
 natbool_10 odd(succ(A8)) -> even(A8)

The theorem is:
 or(and(odd(A9),odd(B0)),and(even(A9),even(B0))) = even(add(A9,B0))
Begin induction on A9
Induction on A9 case 0
 (L) or(and(odd(0),odd(B0)),and(even(0),even(B0)))     << subst A9 with 0 <<
 (R) even(add(0,B0))     << subst A9 with 0 <<
 (L) or(and(false,odd(B0)),and(even(0),even(B0)))     << reduct by natbool_9 <<
 (L) or(false,and(even(0),even(B0)))     << reduct by natbool_2 <<
 (L) and(even(0),even(B0))     << reduct by natbool_4 <<
 (L) and(true,even(B0))     << reduct by natbool_7 <<
 (L) even(B0)     << reduct by natbool_1 <<
 (R) even(B0)     << reduct by natbool_5 <<
 *** equality obtained ***
Induction on A9 case succ(B1)
 (L) or(and(odd(succ(B1)),odd(B0)),and(even(succ(B1)),even(B0)))     << subst A9 with succ(B1) <<
 (R) even(add(succ(B1),B0))     << subst A9 with succ(B1) <<
 (L) or(and(even(B1),odd(B0)),and(even(succ(B1)),even(B0)))     << reduct by natbool_10 <<
 (L) or(and(even(B1),odd(B0)),and(odd(B1),even(B0)))     << reduct by natbool_8 <<
 (R) even(succ(add(B1,B0)))     << reduct by natbool_6 <<
 (R) odd(add(B1,B0))     << reduct by natbool_8 <<
Begin induction on B1
Induction on B1 case 0
 (L) or(and(even(0),odd(B0)),and(odd(0),even(B0)))     << subst B1 with 0 <<
 (R) odd(add(0,B0))     << subst B1 with 0 <<
```

(L) or(and(true,odd(B0)),and(odd(0),even(B0))) << reduct by natbool_7 <<
(L) or(odd(B0),and(odd(0),even(B0))) << reduct by natbool_1 <<
(L) or(odd(B0),and(false,even(B0))) << reduct by natbool_9 <<
(L) or(odd(B0),false) << reduct by natbool_2 <<
(R) odd(B0) << reduct by natbool_5 <<
(L) or(B3,false) << gen of odd(B0) <<
(R) B3 << gen of odd(B0) <<
*** equality depends on next lemma ***
The lemma is:
or(B3,false) = B3
Begin induction on B3
Induction on B3 case false
(L) or(false,false) << subst B3 with false <<
(R) false << subst B3 with false <<
(L) false << reduct by natbool_4 <<
*** equality obtained ***
Induction on B3 case true
(L) or(true,false) << subst B3 with true <<
(R) true << subst B3 with true <<
(L) true << reduct by natbool_3 <<
*** equality obtained ***
End induction on B3
Induction on B1 case succ(B2)
(L) or(and(even(succ(B2)),odd(B0)),and(odd(succ(B2)),even(B0))) << subst B1 with succ(B2) <<
(R) odd(add(succ(B2),B0)) << subst B1 with succ(B2) <<
(L) or(and(odd(B2),odd(B0)),and(odd(succ(B2)),even(B0))) << reduct by natbool_8 <<
(L) or(and(odd(B2),odd(B0)),and(even(B2),even(B0))) << reduct by natbool_10 <<
(R) odd(succ(add(B2,B0))) << reduct by natbool_6 <<
(R) even(add(B2,B0)) << reduct by natbool_10 <<
(L) even(add(B2,B0)) << ind. hyp. on A9 for B2 <<
*** equality obtained ***
End induction on B1
End induction on A9
QED

A Simple General Purpose Technique for Interfacing between Computer Algebra and Numerical Analysis Systems

M. Bayram and J. P. Bennett*

School of Mathematical Sciences
University of Bath
Bath, BA2 7AY
United Kingdom

Abstract. We commonly wish to solve a mathematical problem using a mixture of analytical and numerical techniques. An effective approach is to use a computer algebra system to perform the analytical stage (solution of simultaneous equations, substitution of variables, determination of higher derivatives and so on) and then to use a standard numerical analysis package for the numerical stage (solution of differential equations, fitting to experimental data and so on). Numerical analysis packages are invariably written in languages such as FORTRAN and Pascal, with a rather different input syntax to that of the computer algebra system. To achieve an interface it is necessary for the results from the computer algebra system to be output in a suitable form—FORTRAN expressions for example. Most computer algebra systems now provide fairly sophisticated packages to do this, for example REDUCE's GENTRAN package [Hearn, 1987; Gates, 1987]).
There is a problem with this approach in that the computer algebra system may generate massive expressions (we have a practical example of 4000 terms in 20 variables). On evaluation with specific values in floating point by the numerical analysis package such expressions are prone to serious rounding, overflow and underflow errors. Work is in progress elsewhere to allow expressions to be evaluated minimising such errors. However we demonstrate an alternative approach, where we get the computer algebra system to evaluate the expressions analytically for the numerical analysis package as and when they are needed.
We demonstrate the use of this technique to solve a problem in biochemical kinetics using REDUCE and the NAG FORTRAN library under the Unix operating system. However the technique is completely general and can essentially be applied to any computer algebra system and numerical analysis package working under a multi-tasking operating system.

1 Introduction

We wish to run a FORTRAN numerical analysis program to analyse a very large formula generated by REDUCE. We cannot use conventional systems such as GENTRAN to produce our REDUCE formula in FORTRAN form, since rounding

* To whom correspondence should be addressed. Email J.P.Bennett@bath.ac.uk

and truncation errors in its evaluation in floating point (even double precision) are far too serious.

The solution is to run FORTRAN an REDUCE in parallel, and let FORTRAN call REDUCE whenever it needs the formula evaluated. This can be done very simply under a multi-tasking operating system such as Unix.

2 The Technique

We start by running a REDUCE session in parallel with our FORTRAN numerical analysis program. However we arrange to connect the standard input of each to the standard output of the other. Thus the REDUCE session reads commands that have been written out by the FORTRAN program, and the FORTRAN program can read the results which REDUCE prints out.

Figure 1 shows the code which sets up this system, which for convenience under Unix is written in the C programming language. For simplicity we have stripped out of this code any system error checking. The full code is available by electronic mail from the second author.

We start by creating two pipes. These are like files, but are used for communicating between processes and we need one for each direction. The pipes are a vector of 2 elements each. Element 0 is for reading and element 1 is for writing. We then use **fork()** to create two processes, one running REDUCE, the other a FORTRAN program, **fortprog**. These are started in the two routines **reduce** and **fortran**. To each of these routines we pass one end of each pipe, and these are then mapped on to the standard input and output streams using **dup2**. Finally REDUCE and **fortprog** are started, inheriting these pipes as their standard input and output.

It is convenient to view **fortprog** as the driving routine. It issues commands to REDUCE, and then reads back the result. An example is shown in figure 2 which just calls REDUCE to evaluate an expression. We typically start by bringing in some initialisation code, with the REDUCE command **in "test.red" $**. This will do various initial calculations including defining the expression(s) to be evaluated later. To ensure this command is sent to REDUCE and not buffered by the I/O system we call the system routine **FLUSH**. Finally we call the (user written) routine **NEXTC** to skip past the command prompt which will occur after each REDUCE command. This simply reads in lines until one ending in **:␣** is encountered.

Evaluation of expressions involves issuing a suitable REDUCE command. In the example above we use **sub(x = i, y = j, ratelaw)** with the values of the FORTRAN variables **I** and **J** substituted for **i** and **j**. We again call **FLUSH** to ensure this is sent, **CALL NEXTC** to skip the command prompt and then read in the value REDUCE prints as result.

This may not be the ideal arrangement unless the results are integers. Rational results may be read in as separate numerator and denominator, or the rational may then be evaluated using **ON ROUNDED** to get a floating point result. In this case **PRECISION 15** will prove convenient as corresponding to the maximum precision for 64 bit IEEE floating point numbers.

Finally we can issue a **bye** command to REDUCE to ensure the process shuts down tidily.

```
void  main()
{
        int  atob[2] ;
        int  btoa[2] ;

        pipe( atob ) ;                          /* Create the pipes */
        pipe( btoa ) ;

        if( fork() == 0 )
               reduce( atob[0], btoa[1] ) ;    /* Child runs REDUCE */
        else
               fortran( btoa[0], atob[1] ) ;   /* Parent runs FORTRAN */

}       /* main( void ) */


void reduce( int  data_in,
             int  data_out )
{
        char *argv[] = { "reduce3.4", NULL } ;  /* Argument vector */
        char *envp[] = { NULL } ;               /* Environment vector */

        dup2( data_in, FD_IN ) ;                /* Change file descriptors */
        dup2( data_out, FD_OUT ) ;

        execve( "/usr/local/bin/reduce3.4", argv, envp ) ; /* Run REDUCE */

}       /* reduce( data_in, data_out ) */


void fortran( int  data_in,
              int  data_out )
{
        char *argv[] = { "fortprog", NULL } ;   /* Argument vector */
        char *envp[] = { NULL } ;               /* Environment vector */

        dup2( data_in, FD_IN ) ;                /* Change file descriptors */
        dup2( data_out, FD_OUT ) ;

        execve( "fortprog", argv, envp ) ;      /* Run the FORTRAN program */

}       /* fortran( data_in, data_out ) */
```

Fig. 1. Driver code for the parallel processes.

```
C
C      Initialise REDUCE
C
       WRITE( 6, 99990 )
       CALL FLUSH( 6 )
       CALL NEXTC
C
C      Evaluate an expression and read back the result
C
       WRITE( 6, 99980 ) I, J
       CALL FLUSH( 6 )
C
       CALL NEXTC
       READ( 5, * ) VALUE
C
C      Leave REDUCE tidily
C
       WRITE( 6, 99970 )
       CALL FLUSH( 6 )
C
       STOP
C
99990 FORMAT( 'in "test.red" $' )
99980 FORMAT( 'sub( x = ', I3, ', y = ', I3 ', ratelaw ) ;' )
99970 FORMAT( 'bye ;' )
       END
```

Fig. 2. FORTRAN code to call REDUCE.

3 An Example

We are concerned with estimating the kinetic parameters of systems of enzyme mediated reactions. Our primary interest is in multiple enzyme systems [Bennett, Davenport and Sauro, 1988; Bayram, Bennett and Dewar, 1991], but to illustrate our technique we consider the single enzyme ***malate dehydrogenase*** which catalyses the reaction:

$$\text{oxaloacetate} + \text{NADH} \rightleftharpoons \text{malate} + \text{NAD}^+$$

The behaviour of the enzyme is characterised by a polynomial function or *rate law.* This relates the rate of breakdown of the substrate metabolites (oxaloacetate and NADH here) to the concentrations of the metabolites involved. Such rate laws can be derived from well understood biochemical theory and are written in terms of the concentrations of the metabolites and a number of *kinetic parameters* (k_i) which characterise the particular enzyme.

Using computer algebra we can derive and simplify such individual rate laws. For malate dehydrogenase the rate law is:

$$v = ([E_0] \cdot (k_1 \cdot [\mathrm{NADH}] \cdot k_3 \cdot [\mathrm{oaa}] \cdot k_5 \cdot k_7 - k_4 \cdot k_2 \cdot k_6 \cdot [\mathrm{mal}] \cdot k_8 \cdot [\mathrm{NAD}^+]))/(
k_1 \cdot [\mathrm{NADH}] \cdot k_4 \cdot k_6 \cdot [\mathrm{mal}] + k_1 \cdot [\mathrm{NADH}] \cdot k_4 \cdot k_7
+k_1 \cdot [\mathrm{NADH}] \cdot k_3 \cdot [\mathrm{oaa}] \cdot k_6 \cdot [\mathrm{mal}] + k_1 \cdot [\mathrm{NADH}] \cdot k_3 \cdot [\mathrm{oaa}] \cdot k_5
+k_1 \cdot [\mathrm{NADH}] \cdot k_3 \cdot [\mathrm{oaa}] \cdot k_7 + k_1 \cdot [\mathrm{NADH}] \cdot k_5 \cdot k_7 + k_4 \cdot k_2 \cdot k_6 \cdot [\mathrm{mal}]
+k_4 \cdot k_2 \cdot k_8 \cdot [\mathrm{NAD}^+] + k_4 \cdot k_2 \cdot k_7 + k_4 \cdot k_6 \cdot [mal] \cdot k_8 \cdot [\mathrm{NAD}^+]
+k_2 \cdot k_6 \cdot [\mathrm{mal}] \cdot k_8 \cdot [\mathrm{NAD}^+] + k_2 \cdot k_5 \cdot k_8 \cdot [\mathrm{NAD}^+] + k_2 \cdot k_5 \cdot k_7
+k_3 \cdot [\mathrm{oaa}] \cdot k_6 \cdot [\mathrm{mal}] \cdot k_8 \cdot [\mathrm{NAD}^+] + k_3 \cdot [\mathrm{oaa}] \cdot k_5 \cdot k_8 \cdot [\mathrm{NAD}^+]
+k_3 \cdot [\mathrm{oaa}] \cdot k_5 \cdot k_7)$$

where v is -d[NADH]/dt[2], $[E_0]$ is the total concentration of enzyme in the system and the enzyme is characterised by eight parameters, k_1 to k_8. We consider the system in a closed environment, where there are constraints on the concentrations of all metabolites and by using Gröbner basis techniques we can further simplify this rate law to remove all references to the concentrations of oxaloacetate, NAD^+, and malate.

We have experimental data for this system, in the form of time-course data for the reaction. This provides us with 565 data points of NADH concentration against time as the reaction proceeds [Fisher, 1990]. We wish to fit our overall rate law to this, to obtain estimates of the kinetic parameters k_i.

This is essentially an overconstrained multi-point integration problem. At this stage we resort to numerical techniques and use an established technique to obtain estimates [Bock 1981; Bayram, Bennett and Dewar, 1991]. We use the NAG FORTRAN subroutine library to provide numerical integration and minimisation routines [NAG, 1988]. In summary we provide estimates of k_i, integrate numerically from one experimental data point to the next, and record the difference between the numerical and experimental values as a residual error. We do this for all 565 data points, yielding 564 residual errors. We then use standard least-squares minimisation to obtain better estimates of k_i that will reduce these residual errors.

At numerous points throughout this process we need to substitute values of k_i and NADH concentration into our rate law to get a specific value for the rate. In our initial attempts at this problem we had REDUCE's GENTRAN package produce the rate law in FORTRAN form. However experiments with scaling lead us to suppose that there are significant rounding, overflow and truncation errors occurring in the evaluation of the rate law (which has a large number of terms and involves numbers of wildly different magnitudes). We therefore resort to the technique described above to enable REDUCE to evaluate the expression exactly each time we needed it.

4 Results

We have evaluated this technique using the experimental system described above. To simplify matters we only tried to estimate k_1, giving the remaining k's their published values.

Our first experiment looked at the sum of the squares of the residuals at the end of the first step of the minimisation process (i.e with our initial guess at k_1). We compared three sets of results:

[2] Note the use of the biochemical notation [X] to mean the concentration of X

1. Using a conventional FORTRAN evaluation of the rate law, as produced by GENTRAN.
2. Using REDUCE to evaluate the rate law numerator and denominator exactly, reading them into REDUCE and then carrying out a double precision division.
3. Using REDUCE to evaluate the rate law as a rational number, and then using `on rounded ; precision 15` to express it in REDUCE as a 15 digit floating point number before reading into FORTRAN.

The results are shown in table 1. We see that the results using the combined FORTRAN-REDUCE system are different by a factor of more than 2. There is even a slight variation when the last division is carried out in FORTRAN rather than REDUCE.

Table 1. Sum of squares after first step of minimisation. (a) using GENTRAN to generate all rate law expressions for evaluation in FORTRAN; (b) evaluating numerator and denominator exactly in REDUCE, reading them into FORTRAN and then carrying out a division; (c) evaluating rate law in REDUCE and then using `on rounded ; precision 15` to evaluate it in REDUCE as a floating point number before reading into FORTRAN

	(a)	(b)	(c)
Sum of squares	$1.26014760 \times 10^{-12}$	$2.81358979 \times 10^{-12}$	$2.81363124 \times 10^{-12}$

Our second experiment then compared the performance of the first and third of the above methods in fitting k_1 using just the first 20 experimental data points. An estimate of the error in the determined value of k_1 was obtained using the bootstrap method [Efron, 1980]. The results are shown in table 2. We see that the new technique takes nearly 10 times as long to run. However we get an error estimate that is nearly 10 times smaller, suggesting that the GENTRAN technique has been responsible for introducing errors.

Table 2. Determination of k_1 with bootstrap error estimates by fitting to 20 data points. (a) using GENTRAN to generate all rate law expressions for evaluation in FORTRAN; (b) evaluating the rate law in REDUCE and then using `on rounded ; precision 15` to evaluate it in REDUCE as a floating point number before reading into FORTRAN

	(a)	(b)
User time (seconds)	34.5	210.0
Sys time (seconds)	1.8	141.2
Total	36.3	351.2
k_1	$1.142 \pm 3.42 \times 10^5$	$1.213 \pm 0.41 \times 10^5$
Sum of Squares	3.754×10^{-12}	4.662×10^{-12}

5 Conclusions

We have shown that a relatively simple technique allows us to call REDUCE from FORTRAN to evaluate expressions accurately when necessary. There is a time penalty (a factor of 10), but the technique may yield a commensurate improvement in accuracy.

The technique is general. Although demonstrated under the Unix operating system, the routines used are common to all systems following the POSIX standard. Similar facilities are available in most multi-tasking operating systems. Slight variations would be needed to use a different computer algebra system, or to interface to a different numerical environment, but the basic technique remains the same.

This technique is still at an early stage of development. For example it is a trivial matter to stop reduce producing prompts, and more attention needs to be paid to avoiding deadlock. However we are now in a position where we can tackle our original problem, the kinetics of multiple enzyme systems. We shall doubtless refine our technique and look forward to publishing further reports in the future.

6 Acknowledgements

M Bayram is a research student funded by the Turkish government. The equipment used in developing this paper was originally provided under SERC grant GR/F 66726.

7 References

Bayram, M, Bennett, J P and Dewar, M (1991). Using computer algebra to determine rate constants in biochemistry. 11th conference of the School of Theoretical Biology, Limoges, France, 3–5 June, 1991.

Bennett, J P, Davenport, J H and Sauro, H M (1988). Solution of some equations in biochemistry. *Bath Computer Science Technical Report* 88-12, University of Bath.

Bock, H G (1981). Numerical treatment of inverse problems in chemical reaction kinetics. In *Modelling of chemical reaction systems*, ed. K H Ebert, P Deuflhard and W Jäger, Springer-Verlag, 1981, 102–125.

Efron, B (1979). Bootstrap methods: Another look at the jackknife. *Annals of Statistics* **7**, pp1-26

Fisher, D L (1990). Novel computer techniques in enzyme kinetics. Final year biochemistry project, published as *Bath Computer Science Technical Report*, 90-41, University of Bath.

Gates, B L (1987). *GENTRAN user's manual.* The Rand Corporation, Santa Monica, California.

Hearn, A C (1987). *Reduce User's Manual Version 3.* The Rand Corporation, Santa Monica, CA 90406.

NAG (1988). *NAG FORTRAN Library Manual—Mark 13.* Numerical Algorithms Group, Oxford.

Recurrent Relations and Speed-up of Computations using Computer Algebra Systems

E. V. Zima

Department of Computational Mathematics and Cybernetics
Moscow State University, Moscow, 119899, Russia
e-mail: zima@cs.msu.su

Abstract. Implementation of various numerical methods often needs organisation of computation using complex iterative formulae, i.e. cycles. The time for performing a program on a computer and the necessary storage capacity are largely dependent on form of these formulae. It is desirable to construct computations so that each iterative step might use the results obtained at the previous steps as completely as possible. This implies the association of the given formulae with the recurrent relations, that bring about the same result but economise the arithmetic operations. A special algebraic method has been created that provides for the automatic construction of such recurrent relations. This method and main ways of its use in the algorithms of cycle optimisation, in the computer algebra systems and in algorithms of automatic parallel programs construction are explained. Many examples and programs are given.

1 Systems of Recurrent Relations

Let the function $f(x,i)$, where $x=(x_1,\ldots,x_s)$ is the set of program variables, and i is an integer-valued variable, be specified by the arithmetic expression $F(x,i)$. Also, let m and $n, m \le n$, be given integers, and assume that we have to compute the values of the function for $i=m, m+1,\ldots,n$. Direct evaluation of $f(x,i)$ with the next value of i often not economic, since not all the results of previous computational steps are utilised (some results are simply lost). Hence it is natural to try to find a connection between the next value of the expression $F(x,i)$ or the next values of certain of its subexpression, and the results obtained at previous steps, i.e. to connect with the function $f(x,i)$ certain systems of recurrent relations (SRR). The simplest SRR's are often used by programmers. There are such relations as

$$h\cdot i = h\cdot(i-1)+h, x^i = x^{i-1}\cdot x, i! = (i-1)!\cdot i$$

Systems of recurrent relations of the type

$$f_{t-1}(x,i)=\begin{cases}\varphi_{t-1}(x), & i=m\\ f_{t-1}(x,i-1)\odot_t f_t(x,i-1), & i>m\end{cases} \tag{1}$$

$$t=1,2,\ldots,k; f_k(x,i)=\varphi_k(x)$$

are being considered in [1,2]. Here $\odot_t \in \{+,*\}, x=(x_1,...,x_s)$ and i,m - integer variables. The integer k will be called the depth of SRR (1) and denoted by $Dp(f_0(x,i))$. Note that the depth of system (1) defines the number of operations

$\odot_j$, which must be performed to obtain the next value of $f_0(x,i)$, and the amount of memory, required to store the intermediate results. Such systems are completely determined by the set of functions $\varphi_0(x),\ldots,\varphi_k(x)$ and the values of operations $\odot_1,\ldots,\odot_k$. Therefore, for shorting we shall write such systems in linear form:

$$f_0(x,i) = \{m, \odot_1, \varphi_0(x), \odot_2, \varphi_1(x), \ldots, \odot_k, \varphi_{k-1}(x), \varphi_k(x)\}$$

Algebraic operations are determined upon (1) type systems. Let it be, e.q.,$m = 0$. By using system $\{0, *, x^{-1}, *, x^2, x^4\}$, which conforms to the function $f(i) = x^{2i^2-1}$ and system $\{0, *, 1, *, z^2, *, z^6, z^6\}$, which conforms to the function $g(i) = z^{i^3+i}$, it is easy to get the system of the type (1), which will conform to the function $h(i) = x^{2i^2-1}z^{i^3+i}$. This is system

$$\{0, *, x^{-1}, *, x^2z^2, *, x^4z^6, z^6\}$$

We can write out expression that relate the depths of the available and newly obtained systems:

$$Dp(h(i)) = max(Dp(f(i)), Dp(g(i))).$$

2 Construction of Systems of Recurrent Relations

A general algorithm of construction of systems of type (1) by the expression which sets the function which is calculated in cycle is worked out. The construction is done in endorder circuit of the binary tree which defines this expression. The possibility of this construction based on the fact that any leaf node of this tree can be related to the system of recurrent relations of type (1) (that is, the system $\{m, +, m, 1\}$ if the leaf node contains variable i, or the system $\{m, +, u\}$ if the leaf node contains other variable u (number u)), and on the usage of operations which the internal nodes of the tree contain to the systems we already got.

Example 1. We will get the following sequence of SRR's as the intermediate results of the application of the algorithm to the function $f(i) = (2i+1)! \cdot z^{i^2} x^{2i^2+1} (i = 0, 1, \ldots)$:

$2 = \{0, +, 2\}$
$i = \{0, +, 0, 1\}$
$2i = \{0, +, 2\} \cdot \{0, +, 0, 1\} = \{0, +, 0, 2\}$
$1 = \{0, +, 1\}$
$2i + 1 = \{0, +, 0, 2\} + \{0, +, 1\} = \{0, +, 1, 2\}$
$(2i + 1)! = \{0, +, 1, 2\}! = \{0, *, 1, +, 6, +, 14, 8\}$
$z = \{0, +, z\}$
$i^2 = \{0, +, 0, 1\}^{\{0,+,2\}} = \{0, +, 0, +, 1, 2\}$
$z^{i^2} = \{0, +, z\}^{\{0,+,0,+,1,2\}} = \{0, *, 1, *, z, z^2\}$
$2i^2 = \{0, +, 2\} \cdot \{0, +, 0, +, 1, 2\} = \{0, +, 0, +, 2, 4\}$
$2i^2 + 1 = \{0, +, 0, +, 2, 4\} + \{0, +, 1\} = \{0, +, 1, +, 2, 4\}$
$x = \{0, +, x\}$
$x^{2i^2+1} = \{0, +, x\}^{\{0,+,1,+,2,4\}} = \{0, *, x, *, x^2, x^4\}$
$z^{i^2} x^{2*i^2+1} = \{0, *, 1, *, z, z^2\}\{0, *, x, *, x^2, x^4\} = \{0, *, x, *, zx^2, z^2x^4\}$

Final result of the algorithm may be write out as expression which operands are systems of type (1):

$$\{0,*,1,+,6,+,14,8\} * \{0,*,x,*,zx^2,z^2x^4\} \tag{2}$$

The methods of organisation of calculations in cycles using such expressions are determined. Program for calculation and output of values $f(i)$ for $i = 0,1,\dots,n$ using constructed SRR's runs as follows:

```
f:=1+x; write(f);
a1:=1; a2:=6; a3:=14; a4:=8;
b1:=x; b2:=z*x^2; b3:=z^2*x^4;
for i:=1 step 1 until n do
  begin
    a1:=a1*a2; a2:=a2+a3; a3:=a3+a4;
    b1:=b1*b2; b2:=b2*b3;
    f:=a1*b1; write(f)
  end
```

Here each iteration consists of 4 multiplications and 2 additions.

3 Generalisations of Systems of Recurrent Relations

Two generalisations of (1) type systems are considered. The first is the result of substitution $f_k(x,i-1)$ in (1) for the function $\varphi_k(x)$, which does not depend on i, i.e.

$$f_0(x,i) = \{m,\odot_1,\varphi_0(x),\dots,\odot_k,\varphi_{k-1}(x),f_k(x,i)\} \tag{3}$$

We can easily show that for any expression there exists a certain system of recurrent relations. The number of arithmetic operations in computation of the next value $f_0(x,i)$ using the system of recurrent relations is not greater than the number of arithmetic operations in the direct computation of $f_0(x,i)$.

Let's consider Example 1. With the help of expression (2) it is easy to get the system of the type (3) which will conform to the function $f(i)$:

$$\{0,*,x,\{0,+,6,+,14,8\} * \{0,*,zx^2,z^2x^4\}\}$$

Program for calculation and output values of $f(i)$ using this system runs as follows:

```
f:=x; write(f);
a2:=6; a3:=14; a4:=8;
b2:=z*x^2; b3:=z^2*x^4;
for i:=1 step 1 until n do
  begin ab1:=a2*b2;
    a2:=a2+a3; a3:=a3+a4; b2:=b2*b3;
    f:=f*ab1; write(f)
  end
```

Here each iteration consists of 3 multiplications and 2 additions.

The second generalisation is intended for the construction of recurrent relations that define values of functions $s_0(i) = \sin P_n(i), c_0(i) = \cos P_n(i)$, where $P_n(i)$ is a polynomial of i. Suppose $P_n(i)$ is a polynomial - at that case values of $P_n(i)$ for $i = m, m+1, \ldots$ are defines by system $\{m, +, \xi_0, \ldots, +, \xi_{n-1}, \xi_n\}$. Therefore values of functions $s_0(i), c_0(i)$ for $i = m, m+1, \ldots$ are defines by system

$$s_{t-1}(i) = \begin{cases} \varphi_{t-1}, & i = m \\ s_{t-1}(i-1)c_t(i-1) + c_{t-1}(i-1)s_t(i-1), & i > m \end{cases} \tag{4}$$

$$c_{t-1}(i) = \begin{cases} \psi_{t-1}, & i = m \\ c_{t-1}(i-1)c_t(i-1) - s_{t-1}(i-1)s_t(i-1), & i > m \end{cases}$$

$$t = 1, 2, \ldots, n; s_n(i) = \varphi_n, c_n(i) = \psi_n$$

where $\varphi_t = \sin \xi_t, \psi_t = \cos \xi_t$.

Such relations make it possible to exclude standard trigonometric function designators from cycle bodies. The set of algebraic operations upon SRR's may be extended. It is easy to define the result of multiplication of the (4) type system and the (1) type system

$$f(x, i) = \{m, *, \delta_0(x), \ldots, *, \delta_{k-1}(x), \delta_k(x)\}$$

where $k \leq n$. It will be system of type (4) in which expressions $\sin \xi_j$ and $\cos \xi_j$, for $j = 0, 1, \cdots, k$ replaced by $\sin(\xi_j)\delta_j(x)$ and $\cos(\xi_j)\delta_j(x)$.

Example 2. In the result of SRR's construction for the function

$$f(x) = e^{\frac{-x^2}{2}} \sin(x^3) 2^{x^3+x} (x = ih, i = 0, 1, \ldots)$$

program for calculation and output values $f(x)$ for $x = 0, h, \ldots, ph$ runs as follows:

```
s0:=sin(0); s1:=sin(h^3) *exp((-h^2)/2)*2^(h^3+h);
s2:=sin(6*h^3) *exp(-h^2)*2^(6*h^3); s3:=sin(6*h^3) *2^(6*h^3);
c0:=cos(0); c1:=cos(h^3) *exp((-h^2)/2)*2^(h^3+h);
c2:=cos(6*h^3) *exp(-h^2)*2^(6*h^3); c3:=cos(6*h^3) *2^(6*h^3);
write(s0);
  for i:=1 step 1 until p do
    begin
      w:=s0*c1+c0*s1; c0:=c0*c1-s0*s1; s0:=w;
      w:=s1*c2+c1*s2; c1:=c1*c2-s1*s2; s1:=w;
      w:=s2*c3+c2*s3; c2:=c2*c3-s2*s3; s2:=w;
      write(s0)
    end
```

Here each iteration consists only of 12 multiplications, 3 additions and 3 subtractions.

4 Cycles Optimisation

Main aspects of the application of the given methods for the optimisation of cycle programs are considered. The method of cycle reorganisation using the construction of systems of recurrent relations was developed. In order to use this method for the cycles without control variable one should find the elementary recurrent relation inside the cycle body (in the cycle with a control variable this variable defines such relation). The solution of this problem in the limits of considered method is given. The method was developed for the reorganisation of computation in functions and procedures which are called from cycle bodies. It can be easily shown that such a well known program optimisation methods as code motion and transformation of inductive variables [3] can be obtained as a result of construction of elementary systems of recurrent relations of (1) type (with $k = 0$ and $k = 1$).

5 Computer Algebra Systems

The situation when the numerical solution of a problem on computer is preceded by symbolic transformations in computer algebra system is typical enough. The formulae which we get in the computer algebra systems are usually unwieldy and direct calculations on them are ineffective. Beside, computations with arbitrary precision used in computer algebra systems make arithmetic operations very expensive. Hence, the problem of economising the number of arithmetic operations in this case is still very essential. For example, consider $f(x) = e^{\frac{-x^2}{2}} \sin(x^3) 2^{x^3+x}$ where $x = 0, h, 2h, \ldots, ph, h = \frac{1}{p}$ and REDUCE-3.3 as computer algebra system. Let $T(p) = \frac{Q(p)}{R(p)}$, where $Q(p)$ stands for time of computations of $f(x)$ using recurrent relations system, and $R(p)$ stands for time of direct computations of $f(x)$. After evaluation of the operators

```
on bigfloat, numval; precision 20;
```

we will get the following :

$$T(10) = 0.24, T(20) = 0.16, T(40) = 0.12$$

Algorithms of construction of systems of type (1),(4) were realised in REDUCE-3.3 computer algebra system. After recurrent relations are constructed, effective code for cycle can be generated.

6 Automatic Parallelisation of Cycle Computations

Now-a-days much attention is payed to the algorithms of parallelising computations that are obviously recurrent [4, 5]. At the same time algorithms of parallelising of arithmetic expressions evaluation do not take into account the fact that these expressions can be in a cycle body [4,5]. It is shown that method of recurrent relations systems construction can be efficiently used to automatic parallelisation of computations in cycles. Systems of recurrent relations have linear structure and can be transformed to "parallel" form more easily then arithmetic expressions that have tree structures. Using methods given it is often possible to reorganise computation effectively and obtain better results in parallelisation.

Let's consider for example the system (1) and multiple processors architecture. Let N processors $P_1, \ldots, P_N (N \geq k)$ and the common array $M[0..L](L \geq N)$ are given. If each processor P_j have registers a_j, b_j, c_j, d_j, e_j, then the general scheme of calculation and output of values $f_0(i)$ will be following:

```
M[j]:=φ_j, j=0,1,...,k;
a_j:= M[j-1], j=1,2,...,k;
for i:=m+1 to n do
  begin
    b_j:= M[j], j=1,2,...,k; write(M[0]);
    a_j:=a_j ⊙_j b_j, j=1,2,...,k;
    M[j-1]:=a_j, j=1,2,...,k;
  end;
write(M[0])
```

Here all statements disposing in a single line can be executed simultaneously (the access to different elements of array M is demanded or it isn't demanded at all for their execution). If the output statements are out of consideration then a calculation of the next value of $f_0(i)$ is demanded one time of parallel system's work. This time consists of 3 steps:
- simultaneous loading of the registers b_j;
- simultaneous executing of the operations $\odot_j$;
- simultaneous storing of the registers' a_j contents in memory.

Let's consider the general scheme of calculation using SRR (4). In this case we'll be use two arrays - $S[0..p]$ and $C[0..p]$ being in common memory. Let $N \geq p$, then the calculation can be organised as following:

```
S[j]:=φ_j, C[j]:= ψ_j, j=0,...,p;
a_j:= S[j-1], j=1,...,p;
c_j:= C[j-1], j=1,...,p;
for i:= m+1 to n do
  begin
    b_j:= S[j], j=1,...,p;
    d_j:= C[j], j=1,...,p;
    e_j:=a_j * d_j + c_j * b_j, j=1,...,p;
    c_j:=c_j * d_j - a_j * b_j, j=1,...,p;
    a_j:= e_j, j=1,...,p;
    S[j-1]:= a_j, j=1,...,p;
    C[j-1]:=c_j, j=1,...,p;
  end;
```

In both this and previous cases all statements being in the single line in the cycle can be executed simultaneously.

Example 4. Let's consider expression

$$y(i) = (2i+1)! + \frac{x^{3i^2+i} \cdot (2i^3+3i+1)}{z^{7i}}, (i = 0, 1, \ldots, n)$$

In the result of SRR's construction we get

$$y(i) = u(i) + v(i) \cdot w(i) \tag{5}$$

where
u(i)=$\{0, *, 1, +, 6, +, 14, 8\}$,
v(i)=$\{0, *, 1, *, x^4/z^7, x^6\}$,
w(i)=$\{0, +, 1, +, 5, +, 12, 12\}$.

Let $N > 10$ (N - is a number of processors). The current values $u(i), v(i), w(i)$ can be obtained simultaneously by one-time of the parallel system's work. After that calculation of expression's (5) value is demanded. But we'll come to the organisation of calculation when calculation of the next value $y(i)$ needs one-time parallel system's work. We'll combine the calculation of $y(i)$ with the calculation of $u(i+1)$ and $v(i+2), w(i+2)$ using information independence of the expression's (5) operands. It demands to take out the part of calculation from the cycle. In program we'll use common arrays $U[0..3], V[0..3], W[0..3]$ and common variables T and Y:

```
U[0]:=1; U[1]:=6; U[2]:=14; U[3]:=8;
V[0]:=1; V[1]:=x^4/z^7; V[2]:=x^6;
W[0]:=1; W[1]:=5; W[2]:=12; W[3]:=12; T:=V[0]*W[0];
V[j]:=V[j]*V[j+1], j=0,1; W[j]:=W[j]+W[j+1], j=0,1,2;
a1:=U[0]; a2:=U[1]; a3:=U[2]; a4:=V[0]; a5:=V[1];
a6:=W[0]; a7:=W[1]; a8:=W[2];
for i:=0 to n do
  begin
    a9:=V[0]; a10:=U[0]; b9:=W[0]; b10:=T;
    b1:=U[1]; b2:=U[2]; b3:=U[3]; b4:=V[1]; b5:=V[2];
    b6:=W[1]; b7:=W[2]; b8:=W[3];
    a10:=a10+b10; a9:=a9*b9;
    a1:=a1*b1; a2:=a2+b2; a3:=a3+b3;
    a4:=a4*b4; a5:=a5*b5;
    a6:=a6+b6; a7:=a7+b7; a8:=a8+b8;
    Y:=a10; T:=a9; U[0]:=a1; U[1]:=a2; U[2]:=a3;
    V[0]:=a4; V[1]:=a5; W[0]:=a6; W[1]:=a7; W[2]:=a8
  end;
```

Here a preliminary calculations of $v(0) \cdot w(0)$ and $v(1), w(1)$ are executed. These preliminary calculations resemble the process of conveyer's start. There are follow equalities after each iteration:

Y=y(i), T=v(i+1)w(i+1), U[0]=u(i+1), V[0]=v(i+2), W[0]=w(i+2).

Each iteration (as early) consists of 3 steps of parallel system's work:
- loading of registers;

- simultaneously execution of operations by different processors;
- storing of registers' contents in memory.

Some calculations are unnecessary in this cycle (for example the calculation of $w(n+1), w(n+2)$ etc.). It may give the attemption of indefinite values calculation. That's why the number of iterations must be decreased and only correct calculations must be written out after the cycle. The upper limit of cycle's parameter can be changed to $n-2$ in Example 4 and then we should write out the cycle's body (without statements connected with the registers $a_4, \ldots, a_8, b_4, \ldots, b_8$) and assignment Y:=U[0]+T. Such completion of the cycle resemble the correct halt of conveyer. The number of iterations is taken for realization of the correct halt is equal to the height of the tree corresponding to expression (5).

We considered early that we had the necessary number of processors. Let's consider the general case. Let $y(i)$ is given expression which is transformed into $s(i)$ after SRR's construction. The operands of $s(i)$ are systems of (1),(3),(4) - type. Let R is the number of constructed SRR's, D - the sum of depths of all constructed SRR's, R1 - the number of inside nodes in the tree corresponding to expression $s(i)$. Then $(D+R1)$ is the number of processors which is necessary for organisation of one-time calculations and $(D+R1+R)$ is the size of demanded memory. If $N \geq R1+D$ then it is easy to realize the one-time calculations of the $s(i)$ values. It's analogous to previous reasoning. If $N < D+R1$ then it's easy to write out the cycle which demands $[\frac{D+R1}{N}] = const$ times of parallel system's work.

Let's consider the organisation of vector calculations in the cycles based on the technique or SRR's construction. Let S, C, U, V, W be vector registers of the length L. The statement S:=SHIFT(V) assigns to a vector S the components of a vector V with an one-component-left shift, i.e.

S[1]:=V[2];S[2]:=V[3];...S[L-1]:=V[L];S[L]:=V[L];

The vector calculations corresponding to SRR of the type (1) are organised more simply. Let for example $\odot_t = *, t = 1,2,\ldots,k; k < L$. Then a program for calculation and output of $f_0(i)$ values where $i = m, m+1, \ldots, n$ looks as following

```
U[j+1]:=φ_j, j=0,1,...,k; U[j]:=1, j=k+2,...,L;
write(U[1]);
for i:=m+1 to n do
  begin
    V:=SHIFT(U); U:=U*V; write(U[1])
  end;
```

Here one vector multiplication is executed at each iteration. The calculations corresponding to SRR of the type (3),(4) are simply organised in common case by using masked arithmetic vector operations [6]. Let's return to the Example 4 and use the expression (5). If $L > 11$ then a program of vector calculation of the same values can look like:

```
U[1]:=1; U[2]:=6; U[3]:=14; U[4]:=8;
U[5]:=1; U[6]:=x^4/z^7; U[7]:=x^6;
U[8]:=1; U[9]:=5; U[10]:=12; U[11]:=12;
Madd:='01100001110'; Mmul:='10001100000';
```

```
write(U[1]+U[5]*U[8]);
for i:=1 to n do
  begin
    V:=SHIFT(U);
    U:=VADD(U,V,Madd);
    U:=VMUL(U,V,Mmul);
    write(U[1]+U[5]*U[8])
  end;
```

Here VADD - the function of addition of masked vectors, VMUL - the function of multiplication of masked vectors and variables Madd and Mmul are used as masks. Thus each iteration demands one vector addition and one vector multiplication.

Now consider the function $f(x) = e^{\frac{-x^2}{2}} \sin(x^3) 2^{x^3+x}$ for $x = 0, h, 2h, \ldots, nh$. Let $L > 4$. So the algorithm of computation of $f(x)$ using systems of recurrent relations can look like:

V[1]:=sin(0); V[2]:=$\sin(h^3)e^{\frac{-h^2}{2}}2^{h^3+h}$;
V[3]:=$\sin(6h^3)e^{-h^2}2^{6h^3}$; V[4]:=$sin(6h^3)2^{6h^3}$;
V[5]:=V[6]:= ... :=V[L]:=0;
U[1]:=cos(0); U[2]:=$\cos(h^3)e^{\frac{-h^2}{2}}2^{h^3+h}$;
U[3]:=$\cos(6h^3)e^{-h^2}2^{6h^3}$; U[4]:=$cos(6h^3)2^{6h^3}$;
U[5]:=U[6]:= ... :=U[L]:=1; write(V[1]);

```
for i:=1 step 1 until n do
  begin
    S:=SHIFT( V ); C:=SHIFT( U );
    W:=V*C + U*S;
    U :=U*C - V*S;
    V:=W; write( V[1] )
  end
```

Here each iteration consists of 6 vector arithmetic operations and 5 vector assignments. We can see that number of vector operations remains the same in the case of calculation of values

$$\sin(P_n(x))exp(Q_m(x))2^{R_t(x)}, x = 0, h, 2h, \ldots$$

(P, Q, R are polynomials of x with power n, m and t), if

$$max(n, m, t) < L - 1.$$

A scheme of vector calculations corresponding to systems of type (1),(3) is a particular case of the scheme corresponding to systems of type (4). Therefore the calculations of expressions values with both SRR's of type (1),(3) and SRR's of type (4) as operands is realized by such scheme using masked vector operations. If $D + R \leq L$ then the body of cycle almost will not differ from one considered. If $D + R > L$ then the body of cycle will consist of series of identical steps (their number is equal to $[\frac{D+R}{L}] = const$).

The construction of SRR's and the organisation of parallel computations by means of SRR demands algebraic transformations of initial expressions. Therefore for realization of considered methods the computer algebra systems are more suitable. The methods of parallel programs generation from SRR's of type (1),(4) are implemented in Reduce 3.3 system.

7 References

1. Zima E.V., Automatic Construction of Systems of Recurrence Relations. (in Journal of Computational Mathematics and Mathematical Physics.), Vol.24, N 6, (1984), pp. 193-197.

2. Zima E.V., Transformations of Expressions Associated with Systems of Recursion Relations. (Moscow University Computational Mathematics and Cybernetics, 1985, N 1, pp. 60-66.)

3. Aho A.V., Ullman J.D., The theory of parsing, translation and compiling. Vol. 2: Compiling. (Prentice-Hall, Inc. Englewood Cliffs, N.J. 1973)

4. Hockney R.W., Jesshope C.R. Parallel Computers: arhitecture, programming and algorithms. Aden Hilgen Ltd., Bristol, 1983.

5. Evans, D.J.(ed.), Parallel Processing Systems. (Cambridge University press, 1982).

6. Supercomputers. Class YI Systems, Hardware and Software. S.Fernbach (ed.), North-Holland, 1986.

Design and Implementation of a Retargetable ALDES Compiler

Gábor Simon

Wilhelm-Schickard-Institut
Universität Tübingen, Germany

Abstract. This work is concerned with an efficient and portable implementation of the ALDES language. The extensive measurements showed an improving the implementation efficiency depending on the application with up to 60% (measured on the Sun 3/260) against the FORTRAN based implementation. The compilation time and the size of the generated code are improved too.

1 Introduction

The *AL*gorithmic *DES*cription Language [8] (ALDES) is designed for algebraic and symbolic computation. It is closely related to the algorithm description style used by Knuth. The efficiency of arithmetic, list processing using automatic storage deallocation, and array handling are equally emphasized in the language. The SAC-2 library [1] consists of a large number of algorithms for computer algebra, written in ALDES. The ALDES/SAC-2 system [2], currently distributed, is based on a translator from ALDES to standard FORTRAN, allowing portability.

Until now the ALDES/SAC-2 system has been mostly implemented on mainframe machines or on high-end workstations, using the FORTRAN based ALDES translator. There are a few other implementations too, differing in performance and portability, either generating machine specific assembly code for the Siemens 7000 and IBM 370 machines [10] or they are based on Lisp [7], Modula-2 [6] and C [4].

In this paper we present design and implementation of an ALDES compiler, improving the implementation efficiency compared to the FORTRAN based compiler, and maintaining portability on the target machine level. Beyond efficiency and portability there were two other points motivating us: by reducing the object size of the ALDES/SAC-2 programs we wanted to make implementation of the SAC-2 system on fast microcomputers e.g. m680x0 or i386 processors possible. We were also interested in application dependent code optimization. Most of the commercial compilers allow turning the global optimizer on and off only, giving the user no control over the quality and quantity of the optimization.

To realize these goals we have chosen to use the Amsterdam Compiler Kit (ACK) [11]. ACK is an integrated program collection for producing portable compilers, supporting Algol like languages and byte-addressable target microprocessors, such as vax, pdp, m680x0, and i80x86.

2 The ALDES Language and the Translator

For better understanding of our design strategy, we give an overview of the present ALDES translator. The translator itself, written in ALDES, uses an intermediate language, the so called Elementary ALDES (EA), which is a subset of the full language providing a semantic foundation. The translator has several functional parts, for parsing; generating declarations, expressions, algorithms for EA; handling recursion; doing some optimization; and generating the host language e.g. **FORTRAN**.

Let us consider the following sample ALDES algorithm. The example below is written in publication ALDES, used for documentation of algorithms.

$M \leftarrow$ **CINV**(L)

[Constructive inverse. L is a list. $M = INV(L)$. M is constructed using $COMP$.]

safe a, L'.

(1) $M \leftarrow ()$; $L' \leftarrow L$; *while* $L' \neq ()$ *do* {ADV(L'; a, L'); $M \leftarrow$ COMP(a, M)}; *return* ∎

An algorithm has three basic parts: a header, a specification, and a body. The body contains one or more numbered steps and terminates with the ∎ symbol. *CINV* is a function resulting in list M, which is the non-destructive inverse of the input list L (i.e. L remains unchanged). In order to improve our insight into the semantic of the language, we have a look at the data structures of ALDES.

2.1 Data Structures and their Realization in ALDES

This section does not give a systematic treatment of the ALDES data structures, we only want to summarize some of them. A more detailed description can be found in [8].

The most important data structures of ALDES are integers, characters, strings, symbolic constants, lists and arrays. An *integer constant* A is a so called γ-integer, which means that $|A| < \gamma$, where γ is the largest positive single precision integer on the target machine. *Character constants* are represented by integers, which may be considered as ordinal numbers of a given character set. *String and symbolic constants* are special cases of lists.

Lists plays a primary role in ALDES. They can be considered either as low level objects provided by the host language (e.g. Lisp, see also [7]), or if the host language does not support list processing, lists are considered as high level objects realized by the language capabilities of Elementary ALDES. (cf. section 2.2) A list is a finite sequence of atoms and/or lists, where an atom is a β-integer. a is a β-integer or β-digit if $|a| < \beta$. β is a system implementation parameter. $\beta = 2^{\xi}$ with $\xi \geq 6$, and $3\beta \leq \gamma$. When using a 32 bit processor architecture $\xi = 29$ is a common choice. The empty list () is by definition represented by β, and all other lists are represented by pointer valued terms, pointing to elements of a one-dimensional array **SPACE** realized in EA. The odd indices in this array are called list addresses. To distinguish atoms from list addresses or cell locations the location of the i-th cell is by definition $\beta + i$. The two subfields of a cell are called reductum- and element-field.

ALDES variables can be *declared* as safe, global, intrinsic and as array. *Safe* variables are introduced to improve the efficiency of the automatic storage deallocation. Safe variables denote objects which can be safely ignored by the garbage collector. A variable is *unsafe* if it is a result of the cell allocation function COMP or it is an output parameter of an algorithm which contains unsafe variables. All variables are considered unsafe by default, except input parameters or variables declared explicitly safe. All unsafe variables are referenced through a special stack, implemented in EA using a global array **STACK**. The garbage collector has to concern itself only with the lists which are reachable on the EA **STACK**. Unsafe global variables are maintained as absolute **STACK** references. (**STACK** is also used by recursive algorithms, if the host language does not support recursion.) *Intrinsic variables* are related to some macro expansion facilities. *Arrays* can be static or dynamic, may have one or more dimension. They are declared by an explicit array-declaration or, implicitly, by appending dimension on variables in safe- or global-declarations. An array is called safe if all its elements are safe, otherwise it is unsafe. Dynamic arrays are always unsafe. Passing such an array name as parameter to a subalgorithm, it can be interpreted as the offset of the array base on **STACK**.

2.2 CINV in Elementary ALDES

The internal representation of an elementary ALDES program uses a restricted character set and linear strings. Thus ornamented terms as E.g. $\bar{l}_1^*$ will be mapped to **LLBS1** (where the meaning of the last four characters are: lower case, bar, star, 1). The ornaments are transliterated in clock-wise order. In the next example we present the elementary ALDES version of CINV revealing the realization of lists. Using the aforementioned conventions, L' maps to **LP**, a to **AL** etc.

```
                M:=CINV(L)
       [Subalgorithms IUP, COMP]
       global BETA, BETA1, INDEX, SPACE[10000], STACK[1000].
       safe AL, BETA, BETA1, INDEX, J1Y, L, LP, SPACE, STACK.
(1)    IUP(1).
(2)    STACK[INDEX]:=BETA; LP:=L.
(3)    if LP=BETA then goto 7.
(4)    J1Y:=LP-BETA1; AL:=SPACE[J1Y].
(5)    J1Y:=LP-BETA; LP:=SPACE[J1Y].
(6)    STACK[INDEX]:=COMP(AL, STACK[INDEX]); goto 3.
(7)    M:=STACK[INDEX]; INDEX:=INDEX-1;
       return||
```

In the variables M and L are pointers to the output and input lists, located somewhere in the **SPACE** array. In the declaration part the global and safe data are found, where **BETA**$=\beta$, **BETA1**$=\beta+1$, **INDEX** is the stack pointer of the ALDES stack, called **STACK**, and **SPACE** is the ALDES–heap space, in this case it contains 5000 cells

(10000 words). `J1Y` is a temporary variable. In the following remarks we refer to step numbers of the (EA) algorithm:

Step `(1)` adjust the ALDES stack pointer reserving space for the unsafe local variable M. From this point on, M can be referenced as `STACK(INDEX)`. In step `(2)` M will be initialized as the empty list. Steps `(3)-(6)` realize the while loop of CINV. We note that the subalgorithm ADV (advance) is realized as system intrinsic (steps `(4),(5)`). `LP` represents a list, the cells of which are lying in the `SPACE` field. `LP-BETA` selects the reductum (cdr/rest) and `LP-BETA1` the element (car/first) field of L'. In step `(6)` COMP returns the composition of the object a and the list M to M. Step `(7)` prepares the return value, and releases the ALDES stack frame.

Now consider the effect of the safe declaration in CINV. Lists and pointer valued terms pointing to a list cell have the same representation. L' is in this case a pointer valued term, temporarily points to one or other cell of the input list L. Since all cells of L' are reachable via L, the garbage collector (GC) does not have to care about L'. a is either an atom or a sublist of L and as we will see in the next section, GC can ignore it too. L' and a are placed on the system stack in the stack frame of CINV. Without the *safe* a, L' declaration they would be on the ALDES stack and step `(1)` and `(7)` would contain `IUP(3)` and `INDEX=INDEX-3` respectively.

2.3 Garbage Collection

The garbage collection of ALDES works with a mark and sweep algorithm. All cells available in the array `SPACE` are cascaded into the available cell list $AVAIL$. Each application of the list composition algorithm COMP results in the removal of the first cell from the $AVAIL$ list. When $AVAIL$ becomes empty, COMP calls the GC algorithm, which creates a new $AVAIL$ list containing all the cells in `SPACE`, which are not accessible from `STACK`.

3 Overview of the Amsterdam Compiler Kit

ACK is a compiler building tool written in C under UNIX, containing modular components [11]. The basic idea behind ACK is to use the same intermediate language (EM) for a series of programming languages, generating this common intermediate language using the so called *front end* programs. The final step is to transform the intermediate language to the desired target-machine code, using *back end* programs. To improve the quality of the code, the optimization can be done independently from the source language and target machine on the intermediate language.

To support a new target machine for a language, with existing front end inside the ACK package, one has to build the tables for program generators, generating the back end, target optimizer and assembler. On the other hand to build a compiler for a new language, all one has to do is to write a front end producing EM code and to choose an existing back end.

The intermediate code used by ACK, is the machine language of a simple abstract stack machine EM (*Encoding Machine*).

3.1 The Main Components

- The *front end* builds up a parse tree and uses it for generating reverse Polish like EM code. ACK has front ends for several languages, like Pascal, Modula-2, C, Occam, etc.. Writing a front end, one can choose the word length and address size of the target machine as parameters, and one has to determine the place of the static data structures in the memory, or on the stack.
- The table driven *peephole optimizer* uses about 600 rules to improve the EM code. The table of the pattern replacements contains lines, each of which has a pattern part and a replacement part, both of them composed of a sequence of EM instructions. A typical example is:

  `inc loc adi $3==w` → `loc $2+1 adi w`

  (where `inc` increments the top of stack, `loc` loads a constant onto stack, and `adi` adds two integer on the top of the stack). For example if the word size of the target machine is `w = 4`, then the following replacement can be made:

```
inc                  loc 6
loc 5      →
adi 4                adi 4
```

- ACK has a *global optimization* facility too, doing inter- and intra-procedural and basic block optimizations. However in this paper we do not consider global optimization.
- *Back ends* of the ACK package are generated using a program that is driven by a machine dependent table. This table gives a mapping between the EM code and the target machine's assembly language. The table is compiled along with the back end in advance, for speed considerations.
- The output of a back end is an assembly language algorithm for the target machine. There is a possibility to do additional, machine specific local optimization using a *target optimizer*, which cannot be performed in the machine independent EM peephole optimizer. The algorithm used here is the same as in the EM peephole optimizer. A typical rewrite rule example for the m680x0 processor family is: `add.l #4,sp; move.l X,-(sp)` → `move.l X,(sp)`.
- ACK has its own *assembler* and *linker*. The assembler program is also generated using tables giving the list of instructions, the binary opcode for each of them and the assembly language syntax of the target machine.

3.2 The EM Machine Architecture

The *EM machine* is based on a memory architecture containing a stack for local variables, a static global data area, a heap for dynamic data structures and a distinct address space for instruction. The stack is used for procedure return addresses, actual parameters, local variables and arithmetic operations. The stack grows from high to low addresses and the heap area grows upwards, or if it is necessary upside down. The heap can only be addressed indirectly. The EM memory can be fragmented, which means that it is easily adaptable to various memory architectures.

There are no general purpose registers, but there are a few internal registers with specific functions: PC is a pointer to the next instruction, LB points to the base of the local variables in the current procedure, SP Points to the top of the stack, HP Points to the top of the heap area.

EM instructions have zero or one argument. There are EM instructions to load a variable or constant onto the stack, store the top item on the stack to memory, do arithmetic with the top of stack items, examine the top one or two stack items and branch conditionally, call procedures and return from them. EM has a few more special purpose instructions. Altogether the instruction set contains about 150 instructions. Pseudoinstructions are used to allocate and initialize data storage, and to pass information to other components of the tool kit, e.g. optimizers, and back ends.

The interaction with the environment of an EM program can be done using monitor calls, which are similar to UNIX system calls. A detailed description of the EM machine can be found in [12].

4 Design of the ALDES Front End

4.1 The Parser

We decided to write our front end in ALDES, maintaining the first part of the original translator and as starting-point of the EM code generation we choose Elementary ALDES in a symbolic prefix form. In the next example we present the while loop part of CINV in this internal list form of EA (cf. step (3)-(6) of 2.2):

```
8  (if (ne LP BETA) (goto 4))
5  (set J1Y (dif LP BETA1))
   (set AL (appl SPACE (J1Y)))
6  (set J1Y (dif LP BETA))
   (set LP (appl SPACE (J1Y)))
7  (set (appl STACK (INDEX)) (appl COMP (AL (appl STACK (INDEX)))))
   (goto 8)
```

The only change we made in the previous part of the translator was to turn off the recursion elimination (because of efficiency reasons). There are a few other tasks in this first phase of the translator which could be postponed to the back end of the compiler. For example the elimination of multiple labels etc.

4.2 The Memory Model

To avoid the extra bookkeeping of the ALDES stack used by the original translator, we decide to integrate it on the stack of the EM architecture. In this way the unsafe local variable will be placed into the actual stack frame on the EM stack, a solution which avoids the double indirect addressing of unsafe variables. An alternate place for the unsafe variables would be the heap. It would allow an easier GC implementation, but the arithmetic would become much more complicated. We even allocate the dynamic arrays on the EM stack, because in the assignments in EA it is difficult to differentiate between dynamic and static arrays. These decisions have the consequence that our GC algorithm is more sophisticated than the original one.

Global and local unsafe variables are located on the stack frame of the main program and the actual procedure respectively. Global safe variables (e.g. the list cell container **SPACE**) are allocated in the global data area, and safe locals are also located in the actual procedure's stack frame on top of the unsafe locals.

4.3 Code Generation

As we mentioned earlier, the front end has to carry out the memory allocation for the static data of every subalgorithm. Let us see a part of the EM sequence belonging to the CINV algorithm and a scheme of the stack frame belonging to it:

```
 mes 2,4,4                            ; word_size=4, pointer_size=4
 exp $cinv                                                 |          |
 pro $cinv,20                                              +----------+
 mes 3,0,4,0                          ; L                  |    L     |
 mes 3,-4,4,4                         ; SIW                +----------+
 mes 3,-8,4,4                         ; JOS        LB    ->|   RSB    |
 mes 3,-12,4,0                        ; AL                 +----------+
 mes 3,-16,4,0                        ; J1Y        LB-4    |   SIW    |
 mes 3,-20,4,0                        ; LP                 +----------+
;               ... truncated here ...             LB-8    |   JOS    |
 exa beta                                                  +----------+
beta                                               LB-12   |   AL     |
 bss 4,0,1                                                 +----------+
;                                                  LB-16   |   J1Y    |
 exa space                                                 +----------+
space                                 ;            LB-20   |   LP     |
 bss 40000,0,1                        ; SPACE[10000]       +----------+
;               ... truncated here ...             SP    ->|          |
 loe beta                             ; BETA               +----------+
 stl -8                               ; JOS                |          |
 lol 0                                ; L
 stl -20                              ; LP
;               ... truncated here ...
 lol -8                               ; JOS
 lol -12                              ; AL
 cal $comp
 asp 8
 lfr 4
 stl -8                               ; JOS
;               ... truncated here ...
```

The actual parameter and the locals are referenced on the EM stack using an offset to the LB (Local Base of the current stack frame). The allocation can be done by **mes** pseudoinstructions. Actual parameters, in this case L as input parameter, has non-negative and the locals negative offset. Assuming word size and pointer size 4, the offset -4 (pointing to `SIW`, Stack Information Word) is a place holder for the number of unsafe locals in the current procedure and an indicator which refers to the presence and location of dynamic arrays. In the case of CINV there is no dynamic structure and the number of unsafe variables is one (`JOS`). This means that the EM code generated contains a code fragment: loc 1; stl -4 (push constant 1 to stack; store top of stack to `SIW`). `SIW` plays an important role in the GC algorithm. `SIW` is followed by unsafe locals (here `JOS` alias M) and locals (here a, J1Y and L').

The memory allocation for the global data β, and `SPACE` are realized using `bss` pseudoinstructions.

The parameter passing uses the call-by-value method, on the contrary to the call-by-reference used by `FORTRAN`. The caller pushes the actual parameters onto the stack (using the C convention). In this case, calling COMP from CINV, L and a will be pushed. After returning, the caller pops the parameters (`asp 8`) and loads the function result onto stack (`lfr 4`).

If the current procedure uses one or more dynamic arrays, a further local variable will be introduced (`DIW`, Dynamic Information Word) as the last local allocated at the code generation for the current procedure. `DIW` has to indicate the size of the dynamic arrays in the actual procedure. In the code generation phase, `DIW` is only a place holder, the assignment follows at runtime.

4.4 Garbage Collection

The original GC version has an easy job; using a for loop all elements of the `STACK` array will be visited. If one of the elements is a list head ($> \beta$) it will be marked. In our case the unsafe variables are fragmented on the system stack. To visit them, we have to follow the dynamic chain of the subalgorithms backwards until the LB of the main program is reached. Let us suppose that there are no dynamic structures in use. In this case, one can use a double loop, where the outer one follows the dynamic chain and the inner one visits the unsafe variables in the actual frame, using the information about the number of them contained in the actual `SIW`. Using dynamic arrays the algorithm is similar but somewhat complicated (`DIW` plays a role similar to `SIW`).

It is clear that this "walking" part of the GC algorithm must be more time consuming than the original one, but the general improvements of the EM implementation make up for it. The mark and sweep part works as in the original GC.

The described memory model and the garbage collection have also some drawbacks. The GC operates on the stack only, this means that putting unsafe variables into registers is not allowed. Therefore we have to pass information on the unsafe variables to the back end and the global optimizer to control the register allocation mechanism.

4.5 Implementation Notes

The components of ACK are independent programs in a UNIX pipeline, starting with the ALDES front end and followed by the peephole optimizer, back end, target optimizer, assembler, and linker. The data flow through the pipe is compressed.

In this paper we describe a realization of the compiler on a Sun 3/260 machine. We use peephole and target optimization for the m68020, with some extensions of the original rules. At this time our front end generates human readable EM code, and we are using ACK's *encode* program to compress it in a separate step.

We choose to use the C runtime library for the environmental communication. There are four machine dependent procedures in the ALDES/SAC-2 system which are written in EM (or possible in the target machine's assembler), they are: `clock`, `input`, `output` and the "walking" part of GC.

5 Empirical Results

To see how fast our compiler is, we ran a series of tests on the Sun 3/260, comparing it with the FORTRAN version using the Sun FORTRAN 1.1 compiler, without global optimization. (Using the ALDES translator with the global optimization of the FORTRAN compiler resulted occasionally in 25% longer execution times!)

We now present some results, using source programs computing resultants [5] of two-variable polynomials, and using the subresultant chain algorithm [9].

5.1 Compilation Time and Object Size

The test program we choose for demonstration (`res.ald` containing 88 lines in 9 procedures without the SAC-2 library modules) was compiled using three different compilers. $\mathbf{ald}_{FF}$ uses the ALDES to FORTRAN translator and it was produced with the F77 compiler. $\mathbf{ald}_{EF}$ uses the ALDES to EM front end, and it was produced by $\mathbf{ald}_{FF}$. $\mathbf{ald}_{EE}$ uses the ALDES to EM front end too, but it was produced by $\mathbf{ald}_{EF}$. The resulting compilation times are summed up in the following table:

Compile times of `res.ald` in sec.					Text segment size
Compiler	code generation	compilation	linking	Σ	
$\mathbf{ald}_{FF}$	3.4	9.3	6.1	18.8	6768 byte
$\mathbf{ald}_{EF}$	5.1	5.0	0.5	10.6	3178 byte
$\mathbf{ald}_{EE}$	2.6	5.0	0.5	8.1	3178 byte

Let us note that $\mathbf{ald}_{FF}$ generates FORTRAN, $\mathbf{ald}_{EF}$ and $\mathbf{ald}_{EE}$ generate EM. The generated FORTRAN code has 2581 and the EM code 28352 characters. The text segment size of the resulting object code for this 9 procedures was 54% smaller using $\mathbf{ald}_{EE}$ instead of $\mathbf{ald}_{FF}$.

The compilation time of $\mathbf{ald}_{EE}$ was distributed among the components of the compiler as follows:

Distribution of the execution times in sec.								
front end	encode	opt	cg	top	as	led	cv	Σ
2.6	0.5	0.4	1.8	0.5	1.8	0.5	0.04	8.1

where *encode* is the EM compactifier, *opt* is the peephole-, *top* the target-optimizer, *cg* the code generator, *as* the assembler, *led* the linker and, *cv* converts the ACK object format to the target machine's object format.

5.2 Execution Times

Let us look at the execution times of the polynomial resultant program (**res.ald**), for several polynomial degrees. The polynomials were dense using five bit coefficients and maximum degree n ($n = 3, 4, 5$) in both of the variables x and y. We were using a list space with 10000 cells.

Subresultant chain algorithm				
Degree	# of GC	$\mathbf{ald}_{FF}$ (sec)	$\mathbf{ald}_{EE}$ (sec)	%
3	2	1.31	0.80	60
4	21	9.61	6.18	64
5	165	58.98	39.10	66

The next table shows the results of an example [3] for quantifier elimination using an algorithm by Weispfenning, implemented by R. Loos. The list space contains 40000 cells.

The Davenport-Heintz example			
	$\mathbf{ald}_{FF}$	$\mathbf{ald}_{EE}$	Percent
Time incl. GC	8.1 sec	3.7 sec	45 %
Number of GCs	2	2	
Time of GC	0.53 sec	0.51 sec	
Text segment size	303104 byte	122880 byte	40 %

6 Concluding Remark

Our compiler produces up to 3 times smaller code, and the produced code is 30-75% faster on the Sun 3/260 compared to the **FORTRAN** version. The execution time of the programs compiled with the new compiler was negatively influenced by the number of GC calls, because the GC algorithm can not cope with the average time gain of the other algorithms.

Over all our compiler is faster than the **FORTRAN** solution, despite the fact that generating EM using $\mathbf{ald}_{EF}$ is slower than using $\mathbf{ald}_{FF}$, because the EM source code contains much more characters as the equivalent **FORTRAN** source.

There are further possibilities to improve our results. To get an even shorter compilation time, the front end should generate the compact EM code directly. This dramatically (up to 85%) reduces the size of the front end's output, and as a consequence it makes *encode* superfluous and thus speeds up the writing phase of the front end too.

In order to reduce execution time by improving code quality, one could use the global optimization facilities. To do this one has to modify the register allocation and the live variable analysis part of the global optimizer, to allow the handling of unsafe variables.

Acknowledgments
The author is grateful to Rüdiger Loos for motivation and support. Thanks to Lars Langemyr and Reinhard Bündgen for valuable discussions and comments.

References

1. G.E. Collins, 1980: ALDES and SAC-2 now available. *SIGSAM bull.*, **12**(2):19.
2. G.E. Collins and R.G.K. Loos: SAC-2 system documentation. In Europe available from: R.G.K. Loos, Universität Tübingen D-7400 Tübingen, Germany. In the U.S. available from: G. E. Collins, Ohio State University, Computer Science, Columbus, OH 43210, U.S.A.
3. J.H. Davenport and J. Heintz, 1988: Real Quantifier Elimination is Doubly Exponential. *J. Symbolic Computation*, **5**:29–35.
4. H. Hong, 1990: *An Improvement of the Projection Operator in Cylindrical Algebraic Decomposition.* PhD thesis, Ohio State University, Columbus.
5. D.E. Knuth, 1981: *The Art of Computer Programming, Volume 2 - Seminumerical Algorithms.* Addison-Wesley.
6. H. Kredel, 1989: From SAC–2 to Modula-2. In *Proc. ISSAC'88 Rome, LNCS 358*, pp 447–455, Springer.
7. L. Langemyr, 1986: Converting SAC-2 code to lisp. *SIGSAM bull.*, **14**(1):11–13.
8. R.G.K. Loos, 1976: The algorithm description language ALDES. *SIGSAM bull.*, **14**(4):15–39.
9. R.G.K. Loos, 1982: Generalized polynomial remainder sequences. In B. Buchberger, G.E. Collins, and R.G.K. Loos, editors, *Computer Algebra, Symbolic and Algebraic Computation*, pages 115–137, Springer-Verlag, Wien-New York.
10. M. Sämann, 1978: *ALDES-Code-Erzeugung für Siemens 7.000 und IBM 360/370 Maschinen.* Diplomarbeit. Fakultät für Informatik, Univ. Karlsruhe.
11. A.S.Tanenbaum, H. van Staveren, E.G. Keizer, J.W. Stevenson, 1983: A Practical Toolkit for Making Portable Compilers. *Commun. ACM*, **26**(9):654–660.
12. A.S. Tanenbaum, H. van Staveren, E.G. Keizer, J.W. Stevenson, 1983: Description of a machine architecture for use with block structured languages. Tecnical Report IR–81, Vrije Universiteit, Amsterdam.

Data Representation and In-built Compilation in the Computer Algebra Program FELIX

Joachim Apel* and Uwe Klaus**

Universität Leipzig
Sektion Informatik
Augustusplatz 10–11
D–O–7010 Leipzig
Germany

Abstract. In this paper we discuss implementational aspects of the special-purpose computer algebra system FELIX designed for computations in constructive commutative and non-commutative algebra. Starting with a brief sketch of the underlying data structures and the user programming language we explain principles of the embedded compiler which enables an interactive compilation and linking of user defined algebraic procedures. The implemented evaluation routine makes it possible to mix compiled and interpreted functions. Run time measurements of special algebraic expressions and standard bases computations demonstrate the advantage of the chosen data representation and the difference between compilation and interpretation.

1 Introduction to FELIX

FELIX is specially designed for computations in and with algebraic structures and substructures. The basic domains implemented so far are

- commutative polynomial rings,
- free non-commutative algebras,
- quotient structures (e.g. enveloping algebras of Lie–algebras, algebras of solvable type, and G–algebras), and
- finitely generated modules over the above structures.

Possible coefficient domains can be chosen among

- integers,
- rational numbers,
- complex numbers,
- finite prime fields, and
- rational functions.

FELIX not only manages the calculation with elements of the above algebraic structures but also with the structures themselves and mappings between them as well.

* apel@informatik.uni-leipzig.dbp.de
** klaus@informatik.uni-leipzig.dbp.de

A central part of the system is Buchberger's algorithm for the computation of standard bases. It is necessary for many of the implemented operations over ideals and submodules. Possible ideal operations are

- computation of the reduced standard base,
- sum, product, intersection and quotient of two ideals,
- computation of elimination ideals,
- computation of syzygy modules and syzygy chains,
- free resolutions,
- Hilbert–function, and
- computation of transformation matrices between two ideal bases

[AK91] gives a more detailed overview about the algebraic capabilities of FELIX.

2 Designing Principles of FELIX

Starting-point of the development of FELIX was the implementation of non-commutative algebras as described in [AK90]. An effective calculation with such structures including standard bases computation demands a polynomial arithmetic which supports a wide scope of admissible term orderings and efficient pattern matching algorithms for the critical pair treatment. Using a general-purpose computer algebra system the only possibility to implement most of the required data types is a simulation by lists which has the disadvantage of an increased memory consumption. So, developing FELIX a main goal was an economical data representation. A pleasant side-effect was a diminished computing time as will be shown in section 8.

FELIX consists of three layers. The first one is based on a virtual machine detailed described in [Kl89]. This is always the machine dependent part of the system which provides a host language, mainly for list manipulation. Implementing FELIX means programming this virtual machine. After it the next two layers are invariants.

The second layer realizes the FELIX programming language. The language contains control constructs and a complete procedure concept in a PASCAL–like syntax. Means for syntax definitions of algebraic expressions as usual a mathematican would write down them are also available.

The last layer is a collection of all the in-built algebraic algorithms.

At present FELIX is implemented on IBM–compatible computers (any XT or AT) under the DOS operating system. Unfortunately, the well known restriction of the 640 kByte barrier prevents the computation of serious applications.

To overcome this memory restriction, a second version FELIX386 based on the PharLap DOS extender [P91] was developed. It runs on the Intel processor 80386/486 based computers and uses all the available memory. Beyond, the swapping techniques of the PharLap virtual memory manager allow the allocation of additional memory. Running applications this is effected automatically when the physical memory is exhausted.

In both cases, the above mentioned first layer was programmed in assembler language. The approximate size of the machine code is 35 kByte and 40 kByte, respectively. The concrete size of some parts distinguished in their tasks is shown

Table 1. Size of machine code in kByte

Task	FELIX	FELIX386
list processing	1.9	2.2
garbage collector	3.8	4.4
interpreter	2.2	2.8
compiler / linker	6.0	6.8
input / output	3.3	3.7
rational arithmetic	5.4	5.6

in table 1. The PharLap based implementation is a pure protected mode application and uses the flat linear memory model [I90].

3 Data Types and Memory Management

The structure of the FELIX data is LISP–like. As usual within computer algebra systems, any data is either an atom or a sequence of other data (lists). The classes of atoms implemented so far are

- *names*,
- *integers*,
- *rational numbers*,
- *character strings*,
- *exponent vectors*, and
- *bitstrings*.

In contrast to many LISP–implementations the storage model of FELIX is inhomogeneous, i.e. the different data types are stored in different storage areas (see figure 7).

The interpretation of *names* is context dependent. The same name can be used as a notation of a global variable as well as of an operator. Each name is registered by a name cell which consists of five entries (see figure 1) created by the first appearance of the name.

The implementation of *integers* distinguishes between short and long numbers. Shorts are within the range of a machine word and stored in their cells (see figure 2).

The data type of *exponent vectors* was created to directly support a commutative polynomial arithmetic. It is used for a sparse representation of a monomial exponent vector. Within the first layer of FELIX there are included sixteen machine routines which implementing a polynomial arithmetic efficiently perform most of the required operations.

The *bitstrings* correspond to the non-commutative case. A non-commutative monomial is stored by a sequence of integers which represent the ring indeterminates. These integers are coded with some bits only since usually the number of ring indeterminates is small.

Long integers, *character strings*, *exponent vectors*, and *bitstrings*, which correspond to data of variable size, are represented by two parts: a cell where they are registered (see figure 5), and a heap entry where their elements are stored (see figure 6).

Sequences (lists) are built as binary trees by node cells in a usual way.

pointer at value
pointer at global property[3]
operator definition
notation
hash pointer[4]

Fig. 1. Components of a name cell

hash pointer[4]
short integer

Fig. 2. Components of a short integer cell

pointer at numerator
pointer at denominator

Fig. 3. Components of a rational number cell

A most important fact of such an implemented storage model is that each data, independent from the size, can be represented by a single pointer at the corresponding cell and so fits into processor registers.

There are two different kinds of garbage collection. The first one is caused if no cells are available. It is performed in a usual way. First, beginning with the name

[3] Similar as in LISP.

[4] For fast discovery and unique representation all cells of the same kind are listed in a separate hash table.

[5] Marks end of sequence

pointer at sequence element
pointer at next node cell or NIL[5]

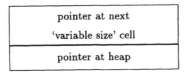

Fig. 4. Components of a node cell

pointer at next 'variable size' cell
pointer at heap

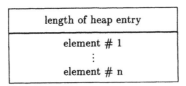

Fig. 5. Components of a cell which represents a variable size object

length of heap entry
element # 1 ⋮ element # n

Fig. 6. Components of a heap entry

cells, the elements on the object stack and the constants of the linked modules, mark all the current occupied data, and than recycle all the unused cells.

The second kind of garbage collection compresses the heap. Note that the heap (see figure 7) is totally movable within the marked area since all the heap entries are chained by the the first parts of all 'variable size' cells. This has the advantage that the situation where a larger, continuous heap entry is requested and an immediately allocation fails, because the heap is plucked to smaller pieces, can be prevented by such a heap compression at any time.

A combined compression and movement of the heap in direction of the linked modules provides for the necessary space to extent the number of cells of any sort. So, cells can be created at that time when they are requested.

The distinction between the object stack, where used FELIX data are located, and program stack, where control data (e.g. return addresses, pointers, counters) are stored, eases the garbage collection since only the object stack has to be treated. The implementation of the necessary stack operations (e.g. PUSH, POP) utilizes the corresponding processor instructions. The switch between both stacks is performed by a single instruction, so that the stack management is very fast.

The performed experiments related with unique data representation have shown that in the case of integers a unique representation of shorts only is a good compromise. It guarantises a good memory exploitation and does not require so much additional computing time to manage the corresponding hash table.

4 Programming Language

The user programming language in FELIX is procedure-oriented and provides the facilities of recursive operator definition and automatic parameter processing.

An operator definition can be assigned to any name. It is registered in the middle part of the corresponding name cell (see figure 1) where four kinds of entries are allowed:

1. address of a subroutine written in machine code: evaluation by an in–built function which performs basic operations with FELIX data (e.g. sum of two integers, total degree of an exponent vector)[6]
2. address of a subroutine within a linked module: evaluation is performed by a compiled function
3. pointer at a sequence whose first element is the key name OPERATOR: this sequence is treated as an operator definition which can be interpreted
4. any other data: no operator definition is bound.

The second layer of FELIX allows the input of operator definitions in a PASCAL–like syntax and transforms them into the above mentioned (3.) internal normal form.

$$\begin{array}{l} \text{OPERATOR} < \mathit{op\_name} > \left\{ \begin{array}{c} (<par\#1>,\ldots,<par\#i>) \\ (\text{NARY}<par>) \end{array} \right\}; \\ \quad \text{LOCALS} < loc\#1 >,\ldots,< loc\#j >; \\ \quad \text{FLUIDS} < fld\#1 >,\ldots,< fld\#k >; \\ \quad < statement\#1 >; \\ \qquad\qquad \vdots \\ \quad < statement\#l >; \\ \text{ENDO} \end{array}$$

The above notation of the formal parameters $< par\ldots >$ determines the arity of the operator bound to the name $< op\_name >$. The arity is fixed to i if the upper notation is used. Otherwise, the key word NARY determines an operator of arbitrary arity.

Noting down the optional LOCALS– or FLUIDS–lines in operator definitions additional local variables $< loc\ldots >$ or $< fld\ldots >$ are available. The difference between both is the visibility.

Locals are only visible within the operator definition where they are introduced (static binding). Their memory location is at the object stack where they are temporary allocated. During transformation of an operator definition into its internal normal form the concrete names of the locals are substituted by place holders.

Fluids are always bound to the value component of the corresponding name cells whose old contents are temporary stored at the object stack. So, their values are also available in other routines called directly or indirectly by this operator definition (dynamic binding).

[6] At present about 160 in–built functions are implemented in the first layer.

Admissible statements are either constants, variables, (algebraic) expressions, or control constructs. Within the first layer of FELIX only the four following control constructs are supported.

The RETURN–statement

```
RETURN (< expr >)
```

finishes the execution of the current operator definition, whose result is the evaluated expression < *expr* >.

The LOOP–statement

```
LOOP < stm#1 >; ... < stm#m >; ENDL
```

causes a cyclic execution of the enclosed statements < *stm* ... >.

The BEGIN–statement

```
BEGIN < stm#1 >; ... < stm#n >; ENDB
```

compounds the enclosed statements < *stm* ... > to a single one.

The WHEN–statement

```
WHEN < expr >;
  < stm#1 >; ... < stm#o >;
ENDW
```

gives the possibility to branch the execution. If the evaluation of the expression < *expr* > yields the Boolean value FALSE the rest of the WHEN–statement is skipped. Otherwise, the above (possibly empty) WHEN–statement sequence < *stm* ... > is carried out. Than the execution of the immediate superordinate structure (operator definition, LOOP–, BEGIN–, or WHEN–statement) is broken off.

These four basic kinds of statements are sufficient to imitate other control construct which usual used in PASCAL–like languages. For instance, the IF–statement

```
IF < expr >
  THEN < stm#i1 >; ... < stm#ip >;
  ELSE < stm#j1 >; ... < stm#jq >;
ENDI
```

can be transformed into the equivalent construction

```
BEGIN   WHEN < expr >;
          < stm#i1 >; ... < stm#ip >;
        ENDW ;
  < stm#j1 >; ... < stm#jq >;
ENDB
```

Such transformations of other control constructs (e.g. FOR–, WHILE–, REPEAT–UNTIL–, CASE–statements) are performed in a similar way within the second layer of FELIX. Since the locals of an operator definition are not bound to names it is easy to create additional, statement–specific local variables (e.g. counter variables in FOR–statements) during parsing. Even in the case of nested statements the number of the required additional local variables can simply be determinated.

5 The Virtual Machine

As already mentioned, the first layer of FELIX realizes a virtual machine whose task is the description of the evaluation of (possibly algebraic) expressions and the data management in a machine–independent way.

It performs an INPUT–EVAL–OUTPUT loop, as usual in interactive computer algebra systems, where the input and output routines are open and can be formed by the user according to his demands. Our ideas (PASCAL–like syntax of operator definitions, usual notation of algebraic expressions, optional TEX–output) are realized within the second layer.

A main task of the virtual machine is the definition of the evaluation (simplification) of any FELIX data object. This is done in the usual recursive way. The evaluation of a variable yields the contents of the corresponding name cell's value component, all other atoms remain unchanged. Evaluating sequences (lists) the first entry is distinguished. In the case of a name, the object is interpreted as an algebraic expression with the first entry's contents as operator name and the remaining entries as arguments (see figure 8). Otherwise, a sequence is evaluated component by component.

In the above case of an algebraic expression the simplification always starts with the evaluation of the arguments $< \mathit{expr} \ldots >$ and then looks into the contents of the middle part of the name cell $< \mathit{op\_name} >$. If the operator is not of fixed arity a sequence of all the evaluated arguments is built and further on treated as the only one argument. In the case of a FELIX in–built function, the arguments are placed into registers and the evaluation is finished by a subroutine call. Otherwise, all arguments are pushed on the object stack where they correspond to the formal parameters. Now, a compiled operator can also be executed by a subroutine call (code generation is described in section 7) and a pure operator definition is interpreted according to the rules of the virtual machine.

Table 2. Use of register

virtual register	task	80x86 register
PP	program stack pointer	(E)SP
OP	object stack pointer	(E)AX
BP	base pointer within the object stack	(E)BP
R0	registers which contain FELIX data objects	(E)SI
R1	registers which contain FELIX data objects	(E)BX
R2	registers which contain FELIX data objects	(E)DI
R3	registers which contain FELIX data objects	(E)DX
RX	general–purpose register for non FELIX data	(E)CX

The virtual machine is register–oriented. There are four data registers (R0, ... , R3) which can contain any FELIX data as described in section 3. The INPUT phase parses an expression, converts it into the internal representation, and puts it into R0. During the EVAL phase the entered expression is simplified and the result is stored again in R0. Finally, the OUTPUT phase prints the evaluated expression. The other data registers are required for the interpretation of operator definition statements. The base pointer BP is used to mark the place on the object stack where the evaluated parameters and local variables of the currently executed operator are located. The both stack pointer (PP, OP) were introduced to manage simply the stack operations. Table 2 summarizes the used virtual registers and their counterparts in the implementation.

6 The Module Concept

The execution of an OPERATOR–statement causes the creation of an operator object and its binding at the corresponding component of the used operator name. An operator object is a list whose first entry is the name OPERATOR. This name refers to an in–built subroutine within the first layer of the system. As any FELIX data object an operator object may be manipulated, i.e. there is no difference between programs and other data.

In particular, all the operators defined within the second and third layer of FELIX were originally created using OPERATOR–statements. At this time the source code of these two layers consists of more than 500 KByte. If one would read in the whole source code at initial time of the system it will cost a lot of time. Of course, it is not done in this way. Actually, the operator definitions are compiled into machine code and stored in so-called module files. During a session the module files may be linked. Linking is much faster than reading in the source code not only because they are only of the half size but, especially, since it is not necessary to modify the data.

Some detailed information about compiler and linker will be given in section 7.

The module concept of FELIX allows to gather operator definitions and to compile them in a common file. On principle, the semantics of interpreted and compiled operators are equal. But creating a module file means more than only to put together a set of different operator definitions. Therefore, it may happen that compiled and interpreted operators behave differently. The operators within a module are strongly connected, i.e. they always prefer to call an operator definition contained in the same module before they try to execute an outside definition.

This will be illustrated by an example. Let a module contain definitions for operators named A and B. After linking the module the operator value of the name B shall be re-defined. If the operator B is called directly the present definition is used. It is impossible to get direct access to the operator definition inside the module. But if the operator A is executed the call of B will cause to execute the definition of B which was valid at compile time of the module, i.e. which is contained in the module file, rather than the present definition of B.

Furthermore, it is even possible to give some of the operator definitions contained in a module the status of help operators, i.e. they are visible only inside the module. There is no other access to a help operator inside a module than to call it by another operator definition compiled in the same module. If a module containing help operators is linked the components of the names associated with the help operators will not be changed.

The module concept has the advantage that calls between operators of one and the same module are very fast. A second advantage is that modules are protected against demaging. A most important fact is that different programmers may restrict their interface to the visible operators which are also called export operators. The multiple use of help operators of the same name in different modules is possible and causes no unwished side effects.

Compiler and interpreter are fully equivalent if any linked module contains only one operator definition, i.e. if the module concept is not applied.

The compiler is called using

$$\text{COMPILE}(< \textit{list\#1} >, < \textit{list\#2} >, \\ < \textit{char\_str\#1} >, < \textit{char\_str\#2} >) \tag{1}$$

The first list consists of the names of the export operators. The second one includes the names of the help operators. The name of the module file to be created is given by the first character string. The last argument serves as a linker message which will be printed during linking. If the compilation is successful the result of the compilation will be the size of the created module otherwise it will be FALSE.

Linking a module can be done using

$$\text{LINK}(< \textit{char\_str} >) \tag{2}$$

The argument is the name of the module file to be linked. The system prevents the user from linking twice the same file, i.e. using the linker inside an operator definition the user needs not to care about whether the file has already been linked before.

An further important feature of the module concept is also that module files may be linked only when they are needed. In section 4 the possible values of the operator component of a name are described. In the current state of the system there will be created an unevaluated operator call in case 4. This will be slightly changed in the future. The system will check in addition whether there exisists a module file which contains an export operator of the wished name. In this case, the module will be linked and the operator executed. Already now, the user needs not care about linking the parts of FELIX since all modules except these for the algebraic structures are linked at initial time and the definition of an active structure causes the linking of the additional necessary modules.

7 In-built Compiler and Linker

The interactive character of the FELIX compiler does not require a repeatedly reading of a source file. When the compiler is called by (1) it is assumed that all operator definitions, which shall be compiled, were already read and converted into their internal normal form.

The compilation starts with a compression of the heap in order to obtain the necessary amount of memory whose use is shown in figure 9.

The tables of export and help operators are of fixed size according to the lenght of the first two input parameters < *list*... >. Their entries contain the operator name and the offset within the generated module where the operator can be called. All used constants and external variables are collected in another table since they are replaced by place holders only. The information about their use is put into the module file (see figure 11), so that they can be properly reconstructed during linking.

The first generated code (briefly sketched in figure 10) is that of all export and help operators. It is always assumed that before calling a compiled operator its arguments were evaluated und placed at the top of the object stack. After allocation of the local variable's entries the current object stack pointer OP is assigned to the base pointer BP. Now, the operator's parameters and local variables can be accessed relative to BP abstracting from their variable names. Since only the parameters and locals of the current executed operator are visible the old value of the base pointer is saved on the program stack. In contrast to such a static binding, the fluid variables are treated individually by accessing the value components of their name cells.

The code of the operator's statement sequence is not immediately generated since its execution can be terminated by a WHEN–statement (see section 4). So, the compilation of any statement sequence is postponed, but replaced by a CALL–instruction with an open destination. The offset of this instruction within the generated module and the sequence are stored on the corresponding stack (see figure 9). After code generation of all operators this stack is reduced entry by entry. Resolving of such an entry means code generation of the sequence in form of a subroutine and updating the invoking CALL–instruction.

In the case of an operator's sequence the subroutine is built in the following way:

```
|op_stm_seq| :  [OLDOP] := OP
                [OLDPP] := PP
                < code_of_stm#1 >
                      ⋮
                < code_of_stm#l >
                RETURN
```

The first two instructions copy the current stack pointers to the memory variables OLDOP and OLDPP. This allows a correct termination of an operator by a RETURN–statement, even if it will occur within a nested statement structure.

The code of the RETURN–statement

```
< code_of_expr >
PP := [OLDPP]
OP := [OLDOP]
RETURN
```

ensures the restoration of these two stack pointers after evaluation of the RETURN–expression.

[7] These PUSH–, POP–, CALL–, RETURN–instructions use the program stack.
[...] means indirect memory access

Because a BEGIN– or LOOP–statement can also be terminated by a WHEN–statement their compilation is performed analogously in two steps. At first, an anonymous CALL–instruction

```
CALL   |begin_stm_seq|     resp.
CALL   |loop_stm_seq|
```

is generated and registered on the stack of statement sequences. In contrast to the OPERATOR–statement a frame is not necessary.

A BEGIN–statement is treated as a subroutine

```
|begin_stm_seq| :   < code_of_stm#1 >
                          :
                    < code_of_stm#n >
                      RETURN
```

when resolving its sequence stack entry.

The code generation of a LOOP–statement sequence ensures the cyclic execution by an unconditional JUMP–instruction:

```
|loop_stm_seq| :   < code_of_stm#1 >
                         :
                   < code_of_stm#m >
                     JUMP |loop_stm_seq|
```

In the case of a WHEN–statement, at first, the code of the expression $< expr >$ is generated. Because the register R0 contains its result at run time, the possible switch is performed by a comparison and a conditional jump instruction:

```
< code_of_expr >
CMP    R0 = FALSE
JNE[8] |when_stm_seq|
```

As usual, the code generation of the WHEN–statement sequence is postponed by creating an corresponding sequence stack entry. Resolving of such an entry means adapting the above jump instruction and the generation of the following code:

```
|when_stm_seq| :   < code_of_stm#1 >
                         :
                   < code_of_stm#o >
                     RETURN
```

The code generation of an (algebraic) expression utilizes its recursive definition

[8] Conditional jump if not equal

(see section 5) and is performed according to its reverse Polish notation. The arguments of any expression (see figure 8) of fixed arity is coded in the following way:

$< \mathit{code\_of\_expr\#r} >$
PUSH R0[9]

$\vdots$

$< \mathit{code\_of\_expr\#1} >$
PUSH R0

In the case of arbitrary arity these instructions are completed by a subroutine call whose execution constructs the sequence of parameters at run time.

The code generation of that expression is finished by instructions which perform the operator evaluation. Operators which are compiled within the module are treated as subroutines:

$$\text{CALL} \quad < \mathit{operator} >$$

where the destination $< \mathit{operator} >$ corresponds to the pattern in figure 10.

The execution of operators outside of the module is performed by

$$\begin{array}{ll} \text{RX} := & r \\ \text{R0} := & < \mathit{op\_name} > \\ \text{CALL} & < \mathit{DISTRIB} > \end{array}$$

At run time a call of $< \mathit{DISTRIB} >$ checks the correct number of parameters and switches according to the current operator definition.

If the expression's operator name $< \mathit{op\_name} >$ corresponds to a FELIX in–built machine routine a direct subroutine call is generated.

Finally, the operator definition's parameters and local variables are coded by a single indexed memory access instruction which allows an access relative to the base pointer BP:

$$\text{R0} := [\ \text{BP} + < \mathit{index} >\ ]$$

Constants, fluid and external variables are replaced by an indirect memory access instruction

$$\text{R0} := [\ \ldots\ ]$$

with an open source operand. During compilation these constants and variable names are collected in tables and reproduced in the module file (see figure 11). The resolving of the open source is part of the linking process.

The code generation is finished when the stack of statement sequences is empty. At the end the compiler routine puts the module file (see figure 11) together.

When the linker is called by (2) the first action is a compression of the heap in order to obtain the necessary amount of memory for the module. Then the module's

[9] These PUSH–instructions use the object stack.

machine code is read in and located at the position which is shown in figure 7. The concrete start addresses of all export operators are entered in the middle part of their name cells. If necessary new name cells are created. Finally, all machine instructions with open operands are resolved.

8 Performances

All computing times presented in the section are given in seconds and measured at an IBM-compatible PC 80386 [10] using FELIX386.

8.1 Number Arithmetic

Table 3 demonstrates computing times for multiplications.

Table 3:
Factorial ($n!$)

n	Time
100	< 0.1
1000	0.6
2000	2.8
3000	6.5
4000	12.0
5000	19.0

Table 4:
Summation ($\sum_{i=1}^{n} 1/i$)

n	Time
100	0.2
200	0.9
300	2.2
500	7.7
800	26.7
1000	49.0
1500	151.0
2000	347.0

Besides multiplication, also GCD-computation and addition are important for the example illustrated in table 4.

8.2 Representation of Commutative Monomials

Commutative monomials are represented using list of their exponents which may be sparse or dense in the system. There is an additional data type for exponent vectors included in the first layer of the system. Such vectors are represented in a packed sparse form. They are manipulated using machine routines.

Table 5 shows the computing times for some Gröbner bases calculations using the different exponent representations. There is computed the reduced Gröbner base of the ideal generated by the polynomials

$$x_1^2 - x_0x_n, x_2^2 - x_1x_n, ..., x_{n-1}^2 - x_{n-2}x_n,$$
$$x_{n-1}x_n$$

with respect to the total degree ordering in the ring $Q[x_0, ..., x_n]$.

[10] 33 MHz (Norton Speed Index: 39.6), 4 MByte RAM

Table 5. Commutative representations

n	**dense**	**sparse**	**exponent vectors**
2	< 0.1	< 0.1	< 0.1
5	1	1	0.5
8	20	15	10
11	185	127	78
14	1112	670	402
17	4632	2626	1571
20	15437	8129	4230

If the number of indeterminates does not exceed five the dense representation is faster than the sparse one. This effect cannot be deduced from the table since the examples are to simple. In any case, the special data type of exponent vectors is always preferable. The other representations are not supported any longer in the current version.

8.3 Representation of Non-commutative Monomials

There will be given computing times for the arithmetic in algebras of solvable type [KW90].

It is possible to represent and compute with non-commutative monomials in several ways. Among these the following are chosen for a computing time comparison presented in table 6.

A most common way of representing products of non-commutative variables is using lists. The calculations are performed by applying rewrite techniques where for each pair of variables there exists a rule turning them into the right order.

A second possibility is to utilize the fact that any element of an algebra of solvable type looks like a commutative polynomial. Only the multiplication procedure has to be changed. In addition, already computed products of two powers of variables may be stored and used as new multiplication rules. This method [Kr90] provides a speed up.

The next two representations are more general in the sense that they are not restricted to algebras of solvable type. They may be used for any quotient structure of tensor algebras.

Starting to implement the system there was used the arithmetic coprocessor for non-commutative computations as multiplication and pattern matching [AK90]. This strategy enables also the treatment of unordered monomials since there is no restriction to the multiplication rules. Monomials had been represented using two 64 bit integers which may be manipulated directly by the coprocessor.

Since this representation was proofed to be rather restrictive to the number of variables and the degree of monomials the bitstrings have been introduced. As table 6 shows the kind of representation is even faster than using the coprocessor. Probably, this effect is due to the rather long communication time between the processors and the requirement of only easy integer operations for which the coprocessor was not intentionally designed.

Table 6. Non-commutative representations

n	**lists**	**solv. type**	**coproc.**	**bitstrings**
4	0.7	0.2	3	0.2
5	3	0.5	10	1
6	8	2	26	3
7	23	4	63	6
8	57	7	134	14
9	135	14	-	32
10	294	25	-	69

Table 6 presents the computing times for calculating $(x+y+z)^n$ in the enveloping algebra of the Lie–algebra $so(3)$ where the multiplication rules are

$$yx = xy - z, zx = xz + y, zy = yz - x.$$

8.4 Interpreter via Compiler

The Gröbner base of the ideal generated by the polynomials
$a + b + c + d + e + f$,
$ab + bc + cd + de + ef + fa$,
$abc + bcd + cde + def + efa + fab$,
$abcd + bcde + cdef + defa + efab + fabc$,
$abcde + bcdef + cdefa + defab + efabc + fabcd$,
$abcdef - 1$
with respect to the degree reverse–lexicographical term ordering should demonstrate the difference between the computing times of interpreter and compiler. The times are given in table 7. The example was detailed discussed e.g. in [D87].

Table 7. Cyclic roots

characteristic	**interpreter**	**compiler**
0	9357	6529
31991	1870	542

Note that in the case of characteristic 0 most of the time is spent for integer arithmetic with large numbers. These arithmetic operations have to be performed using the same machine routines in the interpreted version as well as in the compiled one. The compiler can only reduce the computing time which is necessary for the program control.

In the case of characteristic 31991 the proportion between interpreter and compiler (about 3 : 1) is more favourable since the integer arithmetic is not dominant. For such applications the reached proportion is typical.

8.5 Virtual Memory

The example which should demonstrate the possible computation of standard bases using virtual memory is the so called Big Trinks [BGK86]. The task is to compute the reduced standard base of the ideal generated by

$$45p + 35s - 165b - 36, 35p + 40z + 25t - 27s,$$
$$15w + 25ps + 30z - 18t - 165b^2, -9w + 15pt + 20zs,$$
$$wp + 2zt - 11b^3, 99w - 11bs + 3b^2$$

with respect to the pure lexicographical ordering $b \prec s \prec t \prec z \prec p \prec w$.

Table 8. Big Trinks

physical mem. in MB	virtual mem. in MB	computing time
3	0	14867
0.75	3	41264

The minimal necessary memory for this example amounts 1.5 MByte. The first computation (see table 8) is performed without additional virtual memory. For the second computation the available physical memory was reduced to only the half of the necessary. So, the computing time is trebled.

References

[AK90] J. Apel, U. Klaus, Implementation aspects for non-commutative domains, Proc. IV.Int.Conf.Computer Algebra in Physical Research 1990, JINR Dubna, Moskau, to appear.

[AK91] J. Apel, U. Klaus, FELIX – an assistent for algebraists, Proc. ISSAC'91, ed. S. M. Watt, pp. 382-389, ACM Press, 1991

[BGK86] W. Böge, R. Gebauer, H. Kredel, Some examples for solving systems of algebraic equations by calculating Gröbner bases, J.Sym.Comp., **2**, pp 83-98, 1986.

[D87] J.H. Davenport, Looking at a set of equations, Bath Computer Science Technical Report 87-06.

[I90] Intel Corp., 386^{TM} DX Microprocessor Programmer's reference manual, 1990

[Kl89] U. Klaus, A virtual machine for a computer algebra system (in German), Doctoral thesis, Universität Leipzig, 1989

[Kr90] H. Kredel, Computing in polynomial rings of solvable type, Proc. IV. Int. Conf. Computer Algebra in Physical Research 1990, JINR Dubna, Moskau, to appear.

[KW90] A. Kandri-Rody, V. Weispfenning, Non-commutative Gröbner bases in algebras of solvable type, J.Sym.Comp., **9**, pp 1-26, 1990.

[P91] Phar Lap Software, Inc., 386|DOS–Extender reference manual, 1991

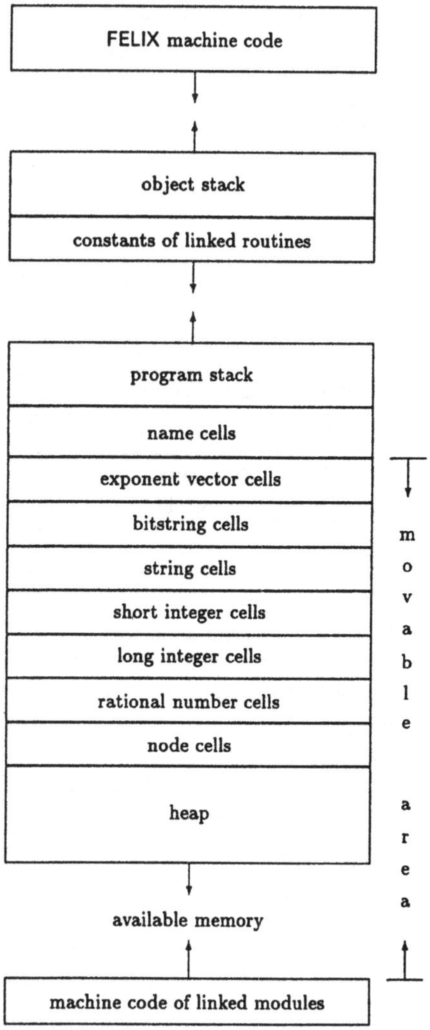

Fig. 7. FELIX386 memory map

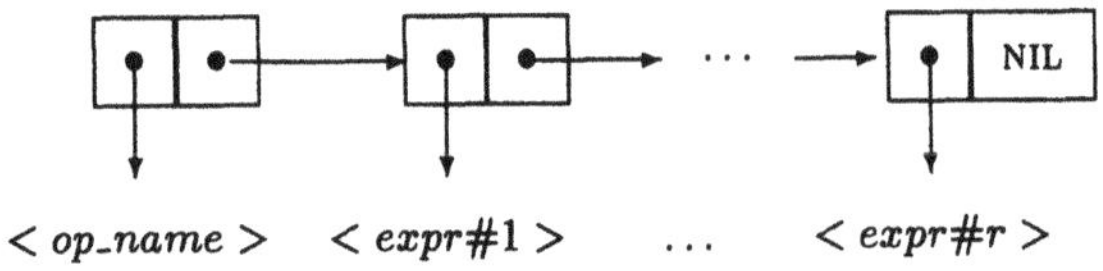

Fig. 8. Representation of an algebraic expression < *op_name* > (< *expr#1* >, ..., < *expr#r* >) using $r + 1$ node cells

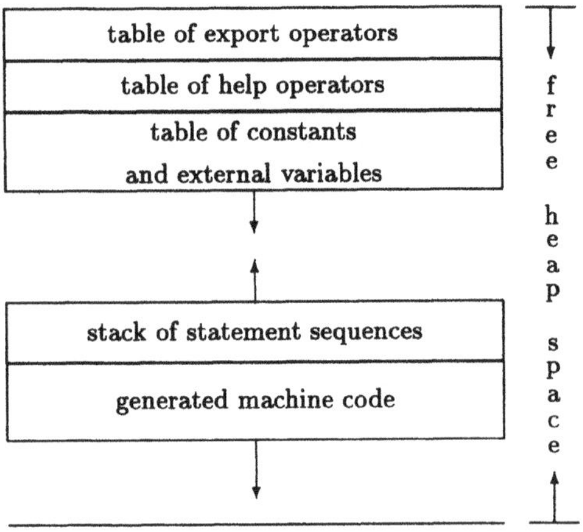

Fig. 9. Use of heap during compilation

check if the number of arguments is correct (for export operators only)
check if program stack overflow occured
check if object stack overflow occured
PUSH [OLDOP][7] PUSH [OLDPP] PUSH BP
allocate j entries on the object stack for the local variables (initialize with FALSE)
BP := OP
for each fluid variable $< fld >$ do: • push value component of the name cell $< fld >$ on the object stack • initialize value component with FALSE
R0 := FALSE ... initial value of operator evaluation
CALL |*op_stm_seq*|
for each fluid variable $< fld >$ (in reverse order) do: • pop element from object stack and restore value component of the name cell $< fld >$
deallocate $i + j$ resp. $1 + j$ entries on the object stack (local variables, parameters)
POP BP POP [OLDPP] POP [OLDOP]
RETURN

Fig. 10. Pattern of operator definition's code

length of machine code
generated machine code
table of export operator names
table of used external variables
table of used constants
linker message

Fig. 11. Scheme of a module file

An Abstract Data Type Development of Graded Rings

A. Goodloe and P. Loustaunau

Department of Mathematics, George Mason University, Fairfax, Va. 22030

1 Introduction

Recently new computer algebra systems such as Scratchpad [10] and Weyl [13] have been developed with built in mechanisms for expressing abstract data types. These systems are object oriented in that they incorporate multiple inheritance and polymorphic types. Davenport and Trager [5] have built much of the framework for basic commutative algebra in Scratchpad II utilizing its' rich set of abstraction mechanisms. In [6] Davenport and Trager concentrated on developing factorization algorithms on domains which were abstract data types.

We are taking a similar approach to the development of algorithms for computing in graded rings. The purpose of this paper is to develop the tools required to compute with polynomials with coefficients in a graded ring $\mathcal{R}$. We focus on graded rings $\mathcal{R}$ which are polynomial rings graded by a monoid, and we allow partial orders on the monomials. The ideas presented here can be applied to more general graded rings $\mathcal{R}$, such as associated graded rings to filtered rings, as long as certain computational "requirements" are satisfied (see Section 5).

Let $\mathcal{R}$ be a Γ-graded ring and let $x_1, \ldots, x_n$ be variables. We assume we have an order on the monomials in $x_1, \ldots, x_n$ given by the ordered monoid $\mathbf{N}^m$. Note that we allow for m to be less than or equal to n, i.e. we allow partial orders (e.g., in the case of two variables x_1, x_2, the order on the monomials $x_1^\alpha x_2^\beta$ could be given by the total degree $\alpha + \beta$, and hence be given by the ordered monoid $\mathbf{N}^1$.) We note that more general orders could be considered, but this does not add anything to the construction presented in this paper. $\mathcal{R}[x_1, \ldots, x_n]$ can then be viewed as a ring graded by $\Gamma \times \mathbf{N}^m$. Two examples are most important in this paper. One is the case $\Gamma = \{0\}$, for the ring $\mathcal{R}[x_1, \ldots, x_n]$ is then the usual polynomial ring, the other is the case $\mathcal{R} = k[y_1, \ldots, y_k]$, $\Gamma = \mathbf{N}^s$, $m = n$ (i.e., we have a term order on the x variables), and we have an elimination order between the x and the y variables. In the latter case, if f is in $\mathcal{R}[x_1, \ldots, x_n]$, then f can be viewed as a polynomial in $x_1, \ldots, x_n$ with coefficients polynomials in $y_1, \ldots, y_k$ ($\Gamma = \{0\}$), or f can be viewed as a polynomial in $y_1, \ldots, y_k, x_1, \ldots, x_n$ with coefficients in k ($\Gamma = \mathbf{N}^s$).

In light of the research that has been done on lifting Gröbner bases from the ring $k[y_1, \ldots, y_k]$ to the ring $k[y_1, \ldots, y_k, x_1, \ldots, x_n]$ (see for example [1, 2, 3, 4, 7, 12]), computing in graded rings has possible applications for the computation of Gröbner bases. Also, there has been some interest in computing Gröbner bases with respect to orders that are not total orders (these bases are then called standard bases), see for example [2, 3]. Moreover, Mora (see for example [11]) has showed that the method of associated graded rings can be an effective tool for computing in power series rings. The natural setting for the latter two computations is the one of graded rings.

In this paper we define the abstract data types for general graded rings $\mathcal{R}$ and their extensions $\mathcal{R}[x_1, \ldots, x_n]$. We use as great a generality as possible with respect to the grading to allow the most flexibility for the user. We will then consider a reduction algorithm similar to the one for polynomials over a field. We are now in the process of implementing the material presented in this paper on the algebra system Weyl.

In [5] Davenport and Trager present a hierarchy of abstract data types for basic commutative algebra. For our purposes it is necessary to add the traditional set inclusion operator, but our hierarchy is the same as in [5].

In order to construct the inclusion operator we must build the operators `insert` and `delete` and also the usual $\in$ operator. These are included in the category `SetModule`. Many definitions for the abstract type `SetModule` may be found in the literature, and we use the one based on the definition in [8]. Throughout the paper, the notation of Scratchpad is used.

2 Graded Ring

We wish to build an abstract data type to represent a Graded Ring. In this paper we will only consider Noetherian rings. As mentioned in [5], there is no useful categorization of the Noetherian condition but we assume that the instances of rings that the user will define will all be Noetherian.

A Graded Ring is defined as an Integral Domain $\mathcal{R}$, where $\mathcal{R}$ is a direct sum of abelian groups:

$$\mathcal{R} = \bigoplus_{\gamma \in \Gamma} R_\gamma$$

where Γ is an ordered monoid. $\mathcal{R}$ has a multiplication operator "$\cdot$" such that

$$R_\gamma \cdot R_\delta \subseteq R_{\gamma+\delta}.$$

The key in formulating the data type is getting an appropriate representation for the notion of direct sum. Notice that $\bigoplus$ can be viewed as appending one element to the next. Hence we represent the graded ring as a list of abelian groups with the append operation :: acting as $\bigoplus$.

Thus the graded ring $\mathcal{R}$ is represented by a list of abelian groups

$$[R_0, R_\gamma, \ldots].$$

Also we rewrite R_γ as the ordered pair (γ, R_γ) using a function

`OrderedAbelianMonoid` $\longrightarrow$ `Domain_of_AbelianGroup`

that assigns to each γ in Γ the corresponding constructor for the particular instance of an abelian group, that is the corresponding direct summand of $\mathcal{R}$. From now on we will write `OAM` for `OrderedAbelianMonoid` and `Domain_of_ AG` for `Domain_of_AbelianGroup`.

In order that the list properly represents the direct sum we need the additional axiom

$$a \in R_\gamma \ \wedge \ a \in R_\beta \Rightarrow \gamma = \beta \ \vee \ a = 0.$$

Before we proceed let us give an abstract formulation of this construction. Each ordered pair in the list is a summand, so we first define a type **Summand** as follows:

$$\texttt{Typedef Summand} == \texttt{OAM} \times \texttt{Domain\_of\_AG}$$

We assume that **Summand** has the usual selectors **first** and **second**.

We define the abstract type **Graded_Ring** as an Integral Domain which is also a list of elements of type **Summand**. We will write **GR** for **Graded_Ring**. More formally:

$$\texttt{Typedef GR} == \texttt{List\_of\_Summand}$$
$$\texttt{where GR inherits IntegralDomain}$$

with appropriate axioms for the multiplication operator "·". We also include the additional axiom that the list be sorted.

We have a constructor to generate the elements of the list.

$$F : \texttt{OAM} \times (\texttt{OAM} \longrightarrow \texttt{Domain\_of\_AG}) \times \$ \longrightarrow \$,$$

where $\texttt{OAM} \longrightarrow \texttt{Domain\_of\_AG}$ is the function that assigns to each $\gamma \in \Gamma$ the abelian group R_γ.

Since F returns a **GR** we may assume that it inserts the newly generated elements in the appropriate place in order to keep the list sorted. Note that F is a higher order function, in that it takes functions

$$\texttt{OAM} \longrightarrow \texttt{Domain\_of\_AG}$$

as parameters. These functions must be defined by the user thus creating an instance of the abstract type. An instance of this function will be denoted by f.

Expressing the above in Scratchpad notation we get:

```
GR(): Category = = Join(IntegralDomain, SetModule,
List(Record(First: OAM, Second: Domain_of_AG))
with
```

$$F : \texttt{OAM} \times (\texttt{OAM} \longrightarrow \texttt{Domain\_of\_AG}) \times \$ \longrightarrow \$$$
$$\cdot : (\texttt{OAM} \times \texttt{Domain\_of\_AG}) \times (\texttt{OAM} \times \texttt{Domain\_of\_AG}) \longrightarrow (\texttt{OAM} \times \texttt{Domain\_of\_AG})$$

Axioms:

$$\begin{aligned}
\texttt{first}(P \cdot Q) &= \texttt{first}(P) + \texttt{first}(Q) \\
\texttt{second}(P \cdot Q) &\subset f(\texttt{first}(P) + \texttt{first}(Q)) \\
&\texttt{SORTED}(\$, <) \\
\texttt{second}(P) \neq 0 &\Rightarrow \texttt{first}(P) \geq 0 \\
\texttt{type(a)} = \texttt{domain(second(P))} &\wedge \texttt{type(a)} = \texttt{domain(second(Q))} \\
&\Rightarrow \texttt{first(P)} = \texttt{first(Q)}
\end{aligned}$$

where $<$ the ordering for **OAM**. The operation **type** uses the system type inference mechanism to infer the type a variable. **domain** returns the name of the domain that the domain constructor is building. **SORTED** is an attribute defined by:

$$\texttt{SORTED}(x :: [\,], <) == \texttt{true}$$
$$\texttt{SORTED}(x :: l, <) == \texttt{if } x \leq \texttt{hd}(l) \texttt{ then true} \wedge \texttt{SORTED}(l, <)$$
$$\texttt{else false}$$

reverse simply reverses the list and **hd** returns the first element of the list.

3 Elements of a Graded Ring

In the previous section we constructed an abstract type for a Graded Ring, now we turn our attention to elements of a graded ring. Recall, given a graded ring

$$\mathcal{R} = \bigoplus_{\gamma \in \Gamma} R_\gamma$$

an element $x \in \mathcal{R}$ is the direct sum of elements of the groups R_γ's,

$$x = \sum_{\gamma \leq \gamma_0} x_\gamma$$

where $x_\gamma \in R_\gamma$.

We define the type **GR_Element** as a pair where the first component is the graded ring domain to which the element belongs and the second is a list of type **Element_Summand**. **Element_Summand** is defined as a pair whose first component is the ordered abelian monoid element indexing the abelian group to which the summand element belongs, and the second component is the summand element itself (which belongs to an abelian group [1]).

Typedef Element_Summand == **OAM_Element** × **AG_Element**

We assume that **Summand** has the usual selectors **first** and **second**. Thus $x = x_\alpha + x_\beta + x_\gamma \ldots$ where $x \in \mathcal{R}$ would look like:

$$(\mathcal{R}, ((\alpha, x_\alpha), (\beta, x_\beta), (\gamma, x_\gamma), \ldots)).$$

We require that the list of summands be sorted. Therefore, for $x \in \mathcal{R}$ as above, we can easily compute the leading term of x, denoted $lt(x)$, and the "value" of x, denoted $\nu(x)$, i.e., the index of the summand $lt(x)$ belongs to. Using the notation above, we have $lt(x) = x_{\gamma_0}$ and $\nu(x) = \gamma_0$.

We build a constructor to generate a list of elements. This constructor takes three parameters as input: an element of some ring which we grade, a function which maps such an element into an element in the ordered abelian monoid thus giving the grading, and an instance of type **GR**. It returns the appropriate instance of **GR_Element**.

G : **Ring_Element** × (**Ring_Element** ⟶ **OAM_Element**) × **Domain_of_GR** ⟶ **$**

Note that the function **Ring_Element** ⟶ **OAM_Element** must be given by the user creating an instance of the abstract type.

Below we give the **GR_Element** type expressed in the Scratchpad notation.

```
GR_Element(): Category = =
```

[1] An element of an abelian group is represented by the abstract type **AG_Element** which is a pair whose first element is the abelian group to which the element belongs and the second is the element itself. The type **OAM_Element** similarly repesents an element of an ordered abelian monoid. We shall assume that simply accessing an element of one of these types yields the element and applying the function **Dom** yields the domain to which the element belongs.

```
Record(Dom: Domain_of_GR, Lst: List (Record (first:
OAM_Element, second: AG_Element ))
with
```

$$G : \texttt{Ring\_Element} \times (\texttt{Ring\_Element} \longrightarrow \texttt{OAM\_Element}) \times \texttt{Domain\_of\_GR} \longrightarrow \$$$

$$\begin{aligned} + &: \$ \times \$ \longrightarrow \$ \\ \cdot &: \$ \times \$ \longrightarrow \$ \\ lt &: \$ \longrightarrow \texttt{AG\_Element} \\ \nu &: \$ \longrightarrow \texttt{OAM\_Element} \\ \texttt{exp} &: \mathbf{N} \times \$ \longrightarrow \$ \end{aligned}$$

Axioms:

$$\begin{aligned}
\texttt{X} + \texttt{0} &= \texttt{X} \\
\texttt{lst_add}(\texttt{x} :: \texttt{X}, [\ ]) &= [\ ] \\
\texttt{lst_add}([\ ], \texttt{y} :: \texttt{Y}) &= [\ ] \\
\texttt{lst_add}(\texttt{x} :: \texttt{X}, \texttt{y} :: \texttt{Y}) &= \texttt{if } (\texttt{first}(\texttt{x}) = \texttt{first}(\texttt{y})) \\
&\quad \texttt{then } (\texttt{first}(\texttt{x}), \texttt{second}(\texttt{x}) + \texttt{second}(\texttt{y})) :: \texttt{lst_add}(\texttt{X}, \texttt{Y})) \\
&\quad \texttt{else if } (\texttt{first}(\texttt{x}) < \texttt{first}(\texttt{y})) \texttt{ then } \texttt{x} :: \texttt{lst_add}(\texttt{X}, \texttt{y} :: \texttt{Y}) \\
&\quad \texttt{else if } (\texttt{first}(\texttt{x}) > \texttt{first}(\texttt{y})) \texttt{ then } \texttt{y} :: \texttt{lst_add}(\texttt{x} :: \texttt{X}, \texttt{Y}) \\
\texttt{Lst}(\texttt{X} + \texttt{Y}) &= \texttt{lst_add}(\texttt{Lst}(\texttt{X}), \texttt{Lst}(\texttt{Y})) \\
\texttt{Dom}(\texttt{X} + \texttt{Y}) = \texttt{Dom}(\texttt{X}) &\wedge \texttt{Dom}(\texttt{X} + \texttt{Y}) = \texttt{Dom}(\texttt{Y}) \\
\texttt{X} \cdot \texttt{0} &= \texttt{0} \\
\texttt{mmult}(\texttt{x}, [\ ]) &= [\ ] \\
\texttt{mmult}(\texttt{x}, \texttt{y} :: \texttt{Y}) &= (\texttt{first}(\texttt{x}) + \texttt{first}(\texttt{y}), \texttt{second}(\texttt{x}) \cdot \texttt{second}(\texttt{y})) :: \texttt{mmult}(\texttt{x}, \texttt{Y}) \\
\texttt{lst_mult}(\texttt{x} :: \texttt{X}, [\ ]) &= [\ ] \\
\texttt{lst_mult}([\ ], \texttt{y} :: \texttt{Y}) &= [\ ] \\
\texttt{lst_mult}(\texttt{x} :: \texttt{X}, \texttt{y} :: \texttt{Y}) &= \texttt{mmult}(\texttt{x}, \texttt{y} :: \texttt{Y}) :: \texttt{lst_mult}(\texttt{X}, \texttt{Y}) \\
\texttt{Lst}(\texttt{X} \cdot \texttt{Y}) &= \texttt{lst_mult}(\texttt{Lst}(\texttt{X}), \texttt{Lst}(\texttt{Y})) \\
\texttt{Dom}(\texttt{X} \cdot \texttt{Y}) = \texttt{Dom}(\texttt{X}) &\wedge \texttt{Dom}(\texttt{X} \cdot \texttt{Y}) = \texttt{Dom}(\texttt{Y})
\end{aligned}$$

$$\begin{aligned}
&\texttt{SORTED}(\$, <) \\
\texttt{lt}(\$) &= \texttt{second}(\texttt{hd}(\texttt{reverse}(\$)) \\
\nu(\$) &= \texttt{first}(\texttt{hd}(\texttt{reverse}(\$)) \\
\texttt{lt}(0) &= 0 \\
\texttt{lt}(1) &= 1 \\
\nu(0) &= 0 \\
\texttt{lt}(\texttt{ab}) &= \texttt{lt}(\texttt{a})\texttt{lt}(\texttt{b}) \\
\nu(ab) &= \nu(a) + \nu(b) \\
\texttt{neg}([\ ]) &= [\ ] \\
\texttt{neg}(\texttt{x} :: \texttt{X}) &= (\texttt{first}(\texttt{x}), -\texttt{second}(\texttt{x})) :: \texttt{neg}(\texttt{X}) \\
\texttt{Lst}(-\texttt{X}) &= \texttt{neg}(\texttt{X}) \\
\texttt{lst_exp}(\texttt{n}, [\ ]) &= [\ ] \\
\texttt{lst_exp}(\texttt{n}, \texttt{x} :: \texttt{X}) &= (\texttt{first}(\texttt{x}), \texttt{exp}(\texttt{n}, \texttt{second}(\texttt{x}))) :: \texttt{lst_exp}(\texttt{X}) \\
\texttt{Lst}(\texttt{exp}(\texttt{n}, \texttt{X})) &= \texttt{lst_exp}(\texttt{X})
\end{aligned}$$

The axioms used in defining addition and multipliplication were built using auxillary functions which define how the operators act upon the list of `Element_Summand`. `lst_add` defines addition of summands having the same index. Multiplication is somewhat more complicated, since each term of the first list is multiplied by each term of the second. The function `lst_mult` takes two summand lists and multiplies each summand in the first by the second list. `mmult` takes a single summand and a list of summands and multplies the first by each term in second.

4 Graded Rings Extensions

Given a ring $\mathcal{R}$, the ring extension $\mathcal{R}[x]$ is simply the ring with elements now taking the form $r_1 + r_2x + \ldots + r_mx^m$, where $r_i \in \mathcal{R}$. So the operators of the ring must be expanded to handle the new variable. Once we define the ring's operators to handle a single variable, it is simple to define them to handle many variables. Most existing computer algebra systems do this for all their pre-defined rings and fields. The problem with the graded ring is the grading, since it must be flexible enough to handle the extra variables.

Let $\mathbf{x} = \{x_1, \ldots, x_n\}$ be variables and $\mathcal{R}$ be a Γ-graded ring. We can order the monomials using the ordered monoid $\mathbf{N}^m$ (we allow partial orders). We wish to construct a new graded ring $\mathcal{R}[\mathbf{x}]$ graded by $\Gamma \times \mathbf{N}^m$ (this new group is ordered by a so-called elimination order, see below). We proceed in two stages:

- We create a function
$$\mathcal{O} : \mathtt{OAM} \longrightarrow \mathtt{OAM}$$
which describes how we order the monomials in $x_1, \ldots, x_n$ (or equivalently, how we order $\mathbf{N}^n$) using $\mathbf{N}^m$. We will denote that order by $\preceq$. The user will define an instance of $\mathcal{O}$ by constructing a function g from $\mathbf{N}^n$ to $\mathbf{N}^m$. The order in $\mathbf{N}^n$ will then be
$$\nu \preceq \nu' \Leftrightarrow g(\nu) < g(\nu') \text{ for } \nu, \nu' \in \mathbf{N}^n.$$
- We then construct the graded ring $\mathcal{R}[\mathbf{x}]$ graded by $\Gamma \times \mathbf{N}^m$. Note that this construction allows the user to consider two very important cases. One is the usual polynomial ring $k[x_1, \ldots, x_n]$ over a field k ($\Gamma = \{0\}$) with "unusual" gradings (such as the total degree grading). The other case is when $\mathcal{R}$ is the polynomial ring $k[y_1, \ldots, y_k]$. Then we can have $\Gamma = \{0\}$ (the coefficients of monomials in $\mathcal{R}[x_1, \ldots, x_n]$ are polynomials in $y_1, \ldots, y_k$) or $\Gamma = \mathbf{N}^k$, where $\mathbf{N}^k$ has a term order (the coefficients are viewed in k, and the elements of $\mathcal{R}[x_1, \ldots, x_n]$ are just polynomials in $k[y_1, \ldots, y_k, x_1, \ldots, x_n]$). As mentioned in the introduction, this has possible application in the theory of Gröbner bases.

The next two subsections expand on these points.

4.1 Building $\mathcal{R}[\mathbf{x}]$ graded by $\Gamma \times \mathbf{N}^m$

As seen at the beginning of this section, given any integral domain $\mathcal{R}$ there is the natural $\mathbf{N}^m$ grading on $\mathcal{R}[\mathbf{x}]$ whose non-zero homogeneous summands are indexed precisely by $\mathbf{N}^m$ using an instance of the function $\mathcal{O}$ above (say g). If $\mathcal{R}$ is graded by Γ (the order is denoted $<_\Gamma$), then let $\Lambda = \Gamma \times \mathbf{N}^m$. We define a Λ grading on $\mathcal{R}[\mathbf{x}]$ as follows: for $\gamma \in \Gamma$ and $\mu \in \mathbf{N}^m$

$$\mathcal{R}[\mathbf{x}]_{(\gamma,\mu)} = \bigoplus_{\nu \in \mathbf{N}^n, g(\nu)=\mu} R_\gamma \mathbf{x}^\nu,$$

where $\mathbf{x}^\nu \equiv x_1^{\nu_1} \cdots x_n^{\nu_n}$ and $\nu = (\nu_1, \ldots, \nu_n) \in \mathbf{N}^n$. We order Λ as follows (the order is denoted $<_\Lambda$):

$$(\gamma_1, \mu_1) <_\Lambda (\gamma_2, \mu_2) \;\Leftrightarrow\; (\mu_1 < \mu_2) \vee (\mu_1 = \mu_2 \wedge \gamma_1 <_\Gamma \gamma_2).$$

This order is called an *elimination order*.

$\mathcal{R}[\mathbf{x}]$ remains a graded ring and thus of the type `List_of_Summand`. In performing the extension we have simply appended $\mathbf{x}$ to R_γ. Thus our list

$$[R_0, R_\gamma, \cdots]$$

becomes

$$\begin{aligned}&[R_0, R_0\mathbf{x}^\nu, \cdots,\\ &\;R_\gamma, R_\gamma\mathbf{x}^\nu, \cdots]\end{aligned}$$

We represent these abelian groups as $n+1$-tuple (R_γ, ν). Note that the addition operation is defined in this new group as follows: for $(r_1, \nu), (r_2, \nu) \in (R_\gamma, \nu)$ as $(r_1 + r_2, \nu)$. The multiplication of two elements in the group is defined as $(r_1, \nu_1) \cdot (r_2, \nu_2) = (r_1 \cdot r_2, \nu_1 + \nu_2)$.

With this theory in place let us outline the procedure for constructing the new graded ring. Initially $\mathcal{R}$ is a Γ-graded ring, represented as a list of ordered pairs, (γ, R_γ). Thus we must transform this list into a new list where for each element, (γ, R_γ), of $\mathcal{R}$, we create a list of elements, $(\gamma, (R_\gamma, \nu))$ of $\mathcal{R}[\mathbf{x}]$ and we impose a new ordering on the graded ring, the elimination order described above.

We generate the summands of $\mathcal{R}$ as we need them using the constructor F. We now build a new constructor, F' for $\Gamma \times \mathbf{N}^m$ from F. The constructor shall be extended automatically using pre-existing objects. This is the content of the next subsection.

4.2 The Ordering on $\mathcal{R}[x_1, \ldots, x_n]$

Given a ring $\mathcal{R}$ graded on Γ and given $\mathbf{N}^m$ we develop the structure which yields the ring $\mathcal{R}[\mathbf{x}]$ graded on $\Gamma \times \mathbf{N}^m$. For that purpose we consider the function

$$\mathcal{H} : \texttt{OAM} \times \texttt{Summand} \longrightarrow \texttt{List\_of\_Summand}.$$

Thus an instance of $\mathcal{H}$ is the mapping:

$$h : \mathbf{N}^m \times R_\gamma \longrightarrow \bigoplus_{\nu \in \mathbf{N}^n} R_\gamma \mathbf{x}^\nu, \text{ where}$$

$$h(\mu, (\gamma, R_\gamma)) = \bigoplus_{\nu \in \mathbf{N}^n, g(\nu)=\mu} R_\gamma x^\nu.$$

Given any fixed $\mu \in \mathbf{N}^m$, it is possible to compute the elements $\nu \in \mathbf{N}^n$ such that $g(\nu_1^i, \ldots, \nu_n^i) = \mu$ (the function g is given constructively). For example, if we were using total degree ordering on 2 variables, and $\mu = 1$, then $\nu = (1,0)$ or $\nu = (0,1)$. Hence we can rewrite the above equation as:

$$h(\mu, (\gamma, R_\gamma)) = [(\gamma, R_\gamma, \nu_1^0, \ldots, \nu_n^0), (\gamma, R_\gamma, \nu_1^1, \ldots, \nu_n^1), \ldots].$$

Each element of the list produced by h is an ordered tuple representing $R_\gamma x_1^{\nu_1} \ldots x_n^{\nu_n}$. In theory the list generated by $\mathcal{H}$ may be infinite, but in our application we will only need a finite number of elements, thus we only generate the finite number of elements in the list that we need. Also note that the new ordered monoid Λ is also well-ordered, so that we can use induction. When new elements are needed they may be built with the constructor G' described below.

To transform our existing graded ring $\mathcal{R}$ into $\mathcal{R}[\mathbf{x}]$ we need simply apply an instance of $\mathcal{H}$ to every element in the list $\mathtt{Lst}(\mathcal{R})$. This gives us a list of lists of type `Summand`. We then apply the `concat` operator to the sublists to obtain a single list. This new list is sorted using the elimination order, $<_\Lambda$, defined above. Expressed in a Miranda[2] like notation we get:

$$\mathtt{sort}((\mathtt{concat} \circ \mathtt{map}\ (h(\mathbf{N}^m, (\mathtt{Lst}(\mathcal{R}))), <_\Lambda).$$

At this stage we have simply extended those elements of $\mathcal{R}$ which are already in the list to $\mathcal{R}[\mathbf{x}]$, we must build new constructor from the old one so that we may add new elements to the list as needed.

Recall that the constructor F is defined as:

$$F : \mathtt{OAM} \times (\mathtt{OAM} \longrightarrow \mathtt{Domain\_of\_AG}) \times \$ \longrightarrow \$$$

We must extend F to F' defined as

$$F' : (\mathtt{OAM} \times \mathtt{OAM}) \times (\mathtt{OAM} \longrightarrow \mathtt{Domain\_of\_AG}) \times \$ \longrightarrow \$$$

where

$$F'(\mu, \nu, f, \mathcal{R}) == h\,(\mu, F(\nu, f, \mathcal{R}))\,.$$

Also, the graded ring element constructor, G , maps an element of a ring R into the ring R graded on Γ. We now need G' which maps an element of $\mathcal{R}[\mathbf{x}]$ into $\mathcal{R}[\mathbf{x}]$ graded on $\Gamma \times \mathbf{N}^m$ An element $x \in \mathcal{R}[\mathbf{x}]$ would look like:

$$(\mathcal{R}[\mathbf{x}], ((\alpha, (x_\alpha, \nu_0)), \ldots, (\beta, ((x_\beta, \nu_0)), ((\alpha, (x_\alpha, \nu_1)), \ldots)).$$

Thus G' simply uses the same function G used,

$$\mathtt{Ring\_Element} \longrightarrow \mathtt{OAM\_Element}$$

to produce a grading on the coefficients of $\mathcal{R}[\mathbf{x}]$. It also maps the coefficient and the exponents of each monomial to the pair forming elements of the group mentioned above. The list of summand elements are sorted using the elimination order.

[2] Miranda is a trademark of Research Software Ltd.

5 Application: Reduction in a Graded Ring

As an application, we present a reduction algorithm. Since we are working over graded rings, we have the concept of leading terms. Namely, let $\mathcal{R}$ be a Γ-graded ring, and let $r \in \mathcal{R}$. then r can be written in a unique fashion as

$$r = \sum_{\gamma \leq \gamma_0} r_\gamma, \text{ where } r_\gamma \in R_\gamma.$$

Now given $r_1, \ldots, r_s$, and $r \in \mathcal{R}$, we can define the reduction of r modulo $r_1, \ldots, r_s$ as follows. If $lt(r)$ is in the ideal generated by $lt(r_1), \ldots, lt(r_s)$, then

$$lt(r) = \sum_{i=1}^{s} a_i lt(r_i).$$

If we consider the element

$$r^* = r - \sum_{i=1}^{s} a_i r_i,$$

then r^* has a leading term strictly smaller than the original $lt(r)$. If $lt(r)$ is not in the ideal generated by $lt(r_1), \ldots, lt(r_s)$, then we consider the element $r^* = r - lt(r)$, and apply the above steps to r^*. This will end in a finite number of steps by the well-ordering condition we imposed on the OAM Γ. The element obtained at the end of the algorithm is called a reduction of r modulo $r_1, \ldots, r_s$.

Note that if $\mathcal{R}$ is the ring $k[x_1, \ldots, x_n]$, where k is a field and the order is a term order, then this is the usual division algorithm, where the ideal membership question raised during the algorithm is easily answered by a test of divisibility of monomials. But the same algorithm can be used for more general graded ring and more general grading, as long as the ideal membership question raised in the algorithm can be answered.

Of more interest, let $\mathcal{R}$ be a Γ-graded ring with a reduction algorithm; i.e. $\mathcal{R}$ is such that an algorithm exists:

1. for deciding whether an element r is in a particular ideal $\langle r_1, \ldots, r_s \rangle$, and if $r \in \langle r_1, \ldots, r_s \rangle$
2. for determining the representation of r as $\sum_{i=1}^{s} a_i r_i$.

(Such rings include polynomial rings, where the algorithm required is the algorithm for Gröbner bases.) Then the reduction algorithm for $\mathcal{R}$ can be easily extended to the ring $\mathcal{R}[x_1, \ldots, x_n]$, graded by $\Lambda = \Gamma \times \mathbf{N}^n$ as in Section 3 (we assume in what follows that the ordering on the monomials in the x variables is a term order; more general orderings can be considered). To reduce a polynomial $f \in \mathcal{R}[x_1, \ldots, x_n]$ by $\{f_1, \ldots, f_k\}$, we first note that the leading term $lt(f)$ of f has two components: a leading coefficient $lc(f)$ in $\mathcal{R}$ (in fact in some R_γ), and a leading monomial $lm(f)$ which is a power product in $x_1, \ldots, x_n$ (recall that we have a term order for the x variables; if the order on the x variables is not a term order, then the leading monomial could be a sum of power products, and we could not separate the leading coefficient from the leading monomial). Thus the membership question raised by the algorithm described at the beginning of the section has now two components: one is whether $lm(f)$ can be divided by some of the $lm(f_i)$ and the other is whether $lc(f)$ is in the ideal generated by the corresponding $lc(f_i)$. Both of these questions can be answered because of the assumption on $\mathcal{R}$.

As mentioned in the introduction this has possible applications in the computation of Gröbner bases.

References

1. W.W. Adams and A. Boyle, "*Some Results on Gröbner Bases over Commutative Rings.*" Journal of Symbolic Computation, (1992) **13**, 473-484.
2. W.W. Adams, A. Boyle, and P. Loustaunau, "*Transitivity for Weak and Strong Gröbner Bases,*" Journal of Symbolic Computation, to appear.
3. W.W. Adams, A. Boyle, and P. Loustaunau, "*An Algorithm for Computing Gröbner Bases,*" to appear in the Proceedings of the Tenth Army Conference on Applied Mathematics and Computing.
4. D. Bayer and M. Stillman, "*A Theorem on Refining Division Orders by the Reverse Lexicographic Order,*" Duke Mathematical Journal, **55** (1987), pp. 321-328.
5. J. H. Davenport and B. M. Trager, "*Scratchpad's View of Algebra I: Basic Commutative Algebra,*" Disco'90, LNCS 429, Springer-Verlag, pp. 40-54, 1990.
6. J. H. Davenport, P. Gianni, and B. M. Trager, "*Scratchpad's View of Algebra II: A Categorical View of Factoization,*" Proceedings of ISSAC'91, ed. S.M. Watt, ACM Press, pp. 32-38 , 1991.
7. P. Gianni, B. Trager, and G. Zacharias, "*Gröbner Basis and Primary Decomposition of Polynomial Ideals,*" Journal of Symbolic Computation, **6** (1988), pp. 148-168.
8. I. Van Horebeek and J. Lewi, "*Algebraic Specifications in Software Engineering,*" Springer-Verlag, 1989.
9. R. Jenks and R. Sutor, "*Type Inference and Coercion Facilities in the Scratchpad II Interpretor,*" SIGPLAN '87 Symposium in Interpreters and Interpretive Techniques, Sigplan Notices, Vol.22, No. 7, ACM Press, pp. 56-63, July, 1987.
10. R. Jenks, R. Sutor and S. Watt "*Scratchpad II: An Abstract Datatype System for Mathematical Computation,*" Trends in Computer Algebra, LNCS 296, Springer-Verlag, 1988, pp. 12-37.
11. T. Mora, "*An Introduction to the Tangent Cone Algorithm,*" preprint.
12. D. Spear, "*A Constructive Approach to Commutative Ring Theory,*" Proceedings 1977 MACSYMA User's Conference, pp. 369-376.
13. R.H. Zippel, "*The Weyl Computer Algebra Substrate Manual,*" Cornell University, 1991.

An Object-Oriented Approach to Indexed Mathematical Objects for the Manipulation of Sums and Series*

Olga Caprotti

Research Institute for Symbolic Computation
Johannes Kepler Universität, A-4040 Linz, Austria
E-mail: ocaprott@risc.uni-linz.ac.at

Abstract. This paper outlines a methodology for the design and implementation of computational symbolic methods that deal with indexed objects, such as variables and functions, appearing in sums and series. We experimented how an integrated design environment, that provides executable object-oriented specifications for axiomatizable mathematical structures, allows to derive a computational system directly from the underlying algebraic properties of the structures to be manipulated. From the formal characterization of indexed objects and of sums and series, we produce the specification of the abstract data types necessary for a correct manipulation of these symbolic objects in CLOS object-oriented environment.

1 Introduction

Indexed variables and functions are used whenever one has to formalize relations in mathematics. Physics, chemistry, and mathematics offer a wide range of applications where expressions containing indexes are involved. Unfortunately, despite being so common, they still cause some problems to the available computer algebra systems as pointed out in [Wan90] and in [CM90]. In this paper, we show how an object-oriented design approach can be applied successfully in order to avoid deficiencies in the manipulation of these symbolic objects.

Systems manipulating indexes often do not recognize bindings and evaluate sums as they were in presence of a constant argument:

$$1: \quad assign\ x_k\ to\ y$$

$$2: \quad \sum_{k=0}^{10} y \longrightarrow 11x_k$$

In a truly symbolic environment, y should be treated as an indexed indeterminate having name x and index k. This idea is naturally conveyed in an object-oriented approach in which y is bound to an object created for representing indexed indeterminates. When attempting to perform the sum with an indexed indeterminate as

*This work has been partially supported by "Progetto Finalizzato Sistemi Informatici e Calcolo Parallelo" of CNR

summand, a method checking both the index of the sum and of the argument gives the right result.

Incorrect binding between the index of the sum and the summand causes ambiguity in case of more complicated summand arguments:

$$\sum_{k=-1}^{10} \int x^k dx \longrightarrow \text{is } k \text{ zero, nonzero or -1?}$$

Such problems can be easily overcome by exploiting the run-time dispatching of methods that is done according to the different possible arguments: we have experimented this feature as it is provided by CLOS [BDG+88].

Furthermore, by knowing the domain of the identifiers, unpreciseness and erroneous operations could be prevented. For example, the methods available to perform differentiation of an indexed expression with respect to an indexed identifier could output different answers depending on the index, and attempting to differentiate with respect to an index indeterminate which is not continuous could be notified. Currently, one can expect the following behavior:

$$\frac{\partial}{\partial x_h}\left(\sum_{k=0}^{10} x_k\right) \longrightarrow 0.$$

Instead, it would be desirable, from a mathematical point of view, to have methods that output either the answer 1, in case $0 \leq h \leq 10$, or the value 0 otherwise.

Up to now, the solving approach has consisted of special-purpose toolkits built upon the various systems. In the case of handling indefinite summations with indexed variables we know the packages [Mio84], [Wan91] and [SWW92] after the need of a correct design specification for manipulating sums and series had been addressed in [Mio83].

As we have experienced, most cases of misbehavior can be related to insufficient type checking of the bound variables appearing in the expression: the decision to apply a procedure does not depend on the actual type of the treated object. These problems encountered in handling summations can be solved if the design of data types and procedures follows the underlying mathematical definitions of indexed indeterminate, of indexed functions and takes into account the properties of summations. An apt methodology is therefore needed to devise such types together with suitable procedures to treat them, not forgetting inherent mathematical properties of the objects to be modeled.

In object-oriented design [LT91], the primary issue is the specification of the classes of objects the system is going to manipulate and this step should be carried out at the abstract level typical of a formal, mathematical description. The abstract data types that are specified can be directly implemented through *classes* that are a combination of modular and typing aspects. *Inheritance*, the principle by which a class may be defined as an extension or a restriction of another, can be fruitfully used when the classes are structured in a hierarchical way. Algebraic domains, like the category hierarchy used in Scratchpad [JT81], are an example of such a situation. *Polymorphism* is the mechanism that permits to have operations referring to objects of more than one class and this can be exploited by providing functions defined abstractly over the proper algebraic domain.

The next section gives a formal characterization of indexed objects, indexed functions and sigma notation with the aim of translating it to a specification of abstract data types. The specification of these types is then summarized in Section 3. Finally, we present some examples of the experiments with executable CLOS code derived from the specification.

2 A Formal Characterization of Indexed Symbolic Objects, Sums and Series

In the following characterization of indexed objects and indexed functions we refer to [Cap91] for the definition of the set of terms **T**, of *indexed function* and for a detailed discussion on the laws of manipulation of sum and series.

2.1 Indexed Objects

When mathematicians use indexes they implicitly take for granted some hypotheses in order for a notation like x_k to be meaningful in a computation. Proceeding by examples, we give a formalization of the intuition behind the use of indexed objects. From the purely syntactical point of view, no restriction can be imposed on indexed notations and they can be used for representing sets or sequences related to the real or rational numbers occurring as indexes, like:

$$S_{\sqrt{2}} = \{i \leq \sqrt{2} : i \in \mathbf{R}\}$$

However, in usual computations an indexed object refers to a correspondence between the domain of the variable and the enumerable set of indexes thus, the index can be assumed to be an integer number and therefore, no computational meaning is assigned to the above notation $S_{\sqrt{2}}$. Nevertheless, it is possible to have more complex indexes such as arithmetical expressions as in x_{k+2*h}. Also, indexes can themselves be indexed as in x_{y_h}, and install a sort of induction in the definition of what is meant by indexed object. We will not be concerned with the case of indexed indexes nor with indexed expressions whose interpretation might be ambiguous, like:

$$\left(x_k + y_h^3\right)_i.$$

However, a correspondence can always be set between such expressions and multiple indexed expressions.

The following characterization of indexed object formalizes all the considerations above. From an abstract point of view, an indexed object reduces to an ordered pair formed by a symbol and an index term: such will be its implementation through the (ADT) abstract data type **Indexed_Ind.** Let $\Gamma = \{a, b, c, \ldots\}$ be a set of constant symbols, $\Lambda = \{x, y, z, \ldots\}$ be a set of symbols, $\Phi = \{\varphi_1, \varphi_2, \varphi_3, \ldots,\}$ be a set of function symbols of arbitrary arity which we will make precise later on, and **T** the set of terms built with these symbols.

Def. 1 *The set of indexed objects* **I** *is defined recursively as follows.*

(i) if $S \in \Lambda$ *is a symbol and* $K \in \mathbf{T}$ *is a term, then the ordered pair* $< S, K > \in \mathbf{I}$ *is the indexed object* S_K.

(ii) if $S \in \mathbf{I}$ *is an indexed object and* $K \in \mathbf{T}$ *is a term, then the ordered pair* $< S, K > \in \mathbf{I}$ *is the indexed object* S_K.

(iii) Nothing else is an indexed object.

In a safe computing environment, some unpleasant situations like x_x should be avoided, and therefore, a first important restriction is that any symbol used as a name of an indexed object cannot occur inside the same object as subterm of the index term. The second restriction comes from the consideration that indexing a variable for computing purposes is, essentially, a way of enumerating elements in a particular domain. In other words, the assignment function which usually, to a simple variable, associates an element of a given domain, in the case of an indexed variable is a function depending on two arguments: the indexed variable and its index. According to the semantic of enumerations, the index should have an integer value, thus the object occurring as index will range only over the integers and the set Φ will be limited to the set $\{+,-\}$.

By a similar approach, we suggest that what is usually intended by indexed function be identified with an indexed object S_K in the following sense. If $f_i(x)$ is an indexed function whose name is f_i and whose argument is x, then the natural way to fit it into the given definition, is by saying that this function's name is the indexed object $< f, i >$ and x is its argument.

Observe that indexed functions having indexed argument are also allowed. The same restrictions that hold between the symbols used as names and those used as indexes in an indexed variable, must also hold between the name of the function and its argument. That is, we do not want situations like $y_x(y)$.

2.2 Sums and Series

We now turn to the definition of a Sigma-notation, which was introduced by J. Lagrange in 1772:

$$\sum_{k:\Delta(k)} A(k)$$

where $A(k)$ is an expression (algebraically is a ring element), Δ is a relation over an ordered set and $k : \Delta(k)$ is the set of all k's satisfying Δ. Our attention will be restricted to sums and to series of *ring elements*. In a sum, A is called the *summand*, and commonly, is an expression containing k that is called the *index* of the sum. A sum is completely specified through the summand and the relation over the index. Straightforwardly from this, one can derive the definition of the ADT **Sigma** of sum objects to be manipulated. Among all possible relations we must distinguish the finite and the infinite case in order to explain the Sigma notation. When $\Delta(k)$ defines a finite set $\{k_1,\ldots,k_n\}$, the Sigma-notation represents the iterated sum:

$$\sum_{k:\Delta(k)} A(k) = A(k_1) + \cdots + A(k_n).$$

Note that, when $\Delta(k)$ reduces to the empty set, the sum is defined to be equal to zero. Typical examples of this kind of relations are those describing intervals as: $0 \leq j \leq 11$.

When $\Delta(k)$ describes an infinite set $\{\ldots,k_{-2},k_{-1},k_0,k_1,k_2,\ldots\}$, the Sigma-notation describes a series and its meaning is related to the behaviour of the limit of the succession of its partial sums. When this limit does not exist, the infinite sum does not exist and the series is called a "divergent" series.

2.3 Laws of Manipulation of Sums and Series

Rather than the conventional evaluation of sums and series as it is done in most computer algebra systems, our aim is to define a set of rules to transform them by manipulating either the summand or the relation over the index. Successfully manipulating sums consists in transforming Sigma-notations in such a way that they become closer to one's goal up to the closed form evaluation. Manipulation rules are generally given to deal with finite sums and, in case of convergent series, some of them can be extended also to infinite sums.

There are four rules that can be considered as basic rules to be applied when dealing with finite sums over a set of integers. The following law, for instance, gives a rule for interchanging the order of summations:

$$\sum_{k:\Delta(k)} \left(\sum_{h:\Theta(h)} A(k,h) \right) = \sum_{h:\Theta(h)} \left(\sum_{k:\Delta(k)} A(k,h) \right)$$

The above is the unspecialized rule that applies when the indexes are unrelated. It can be seen as a subcase of the general rule:

$$\sum_{k:\Delta(k)} \nabla(A(k)) = \nabla \left(\sum_{k:\Delta(k)} A(k) \right)$$

where ∇ is any linear operator. Abstractness requires the specification of this rule in this very general form. Subsequently, the implementation will produce different methods according to the different special cases that depend on the linear operator and on the relation between the indexes. In case of related indexes, that is when the relation $\Theta(h)$ depends also on k, the relations over the indexes are transformed and the system must be able to perform for instance, interval arithmetic. The rule of associativity, which is valid also for convergent series, is a special case of the above rule, too.

Another rule is the distributive law for product of sums which can also be read as a law of factorization for products of sums and it is valid for arbitrary infinite series:

$$\left(\sum_{k:\Delta(k)} A(k) \right) \left(\sum_{h:\Theta(h)} B(h) \right) = \sum_{k:\Delta(k)} \left(\sum_{h:\Theta(h)} A(k)\, B(h) \right).$$

The third rule manipulates the domain of the sum by dividing the domains into subdomains and, conversely, combining several domains into a single one:

$$\sum_{k:\Delta(k)} A(k) + \sum_{k:\Theta(k)} A(k) = \sum_{k:\Delta(k)\vee\Theta(k)} A(k) + \sum_{k:\Delta(k)\wedge\Theta(k)} A(k).$$

No restriction is imposed and it is valid also for any arbitrary infinite series.

The last is the rule for the change of the index variable:

$$\sum_{h:\Delta(h)} A(h) = \sum_{k:\Delta(k)} A(k)$$

the index k of a summand can be considered *bound* to the sigma sign, therefore it is unrelated to other occurrences of k outside the sigma-notation and, under certain conditions, it can be renamed. Shifting of the index can also be derived as a special case of this rule.

Performing a sigma-sum manipulation will involve, in an object-oriented implementation, different methods according to the different types of summands. Among them it will be possible to have again a sum or a linear operator: in both cases a specially designed method will implement properties of sigma which are needed, like interchanging the order of summation for related/unrelated indexes and interchanging the operators.

3 An Object-Oriented Specification of Indexed Symbolic Objects, Sums and Series

The following specification of a system intended for the manipulation of indexed mathematical objects in sums and series translates the formalization given above. The specification language we use is tailored on the object-oriented view of software design and, besides the standard keywords sorts, operations and axioms, we allow, after the keyword inherit, the specification of inherited data types. Types are inherited together with operations which are usually renamed, as is typically the case with accessor functions, otherwise, when redefined in the new type, the new definition hides the inherited one. The keyword proper introduces operations that are defined for the first time for that type. Furthermore, we use parametric specifications in order to express polymorphic data types.

The simplest data type is the **Identifier** type that associates to a symbol its name. It is the supertype of the parametric types **Constant(Domain)**, **Indeterminate(Domain)** and **Indexed_Ind(Domain)**. Data belonging to these types specify variables and constants as they are used in arithmetical expressions: therefore, they are symbols to which attributes like a value and a domain can be associated. Functions that access the attributes name, domain and value are provided; a constructor function is also furnished whose name is suffixed by _c.

Identifier =
    sorts   symbol, identifier, boolean;
    operations
    proper
        identifier_c: symbol → identifier
        name: identifier → symbol
        equal?: identifier identifier → boolean;
    axioms   ∀*id*: identifier,
        (identifier_c(name(*id*)) = *id*);

Domain describes the domain of definition of the identifier and enriches the parametrized type with properties derivable from the algebraic structure of the domain. For instance, an indeterminate can be of type:

Integer_Ind = **Indeterminate**(Integer_Domain);

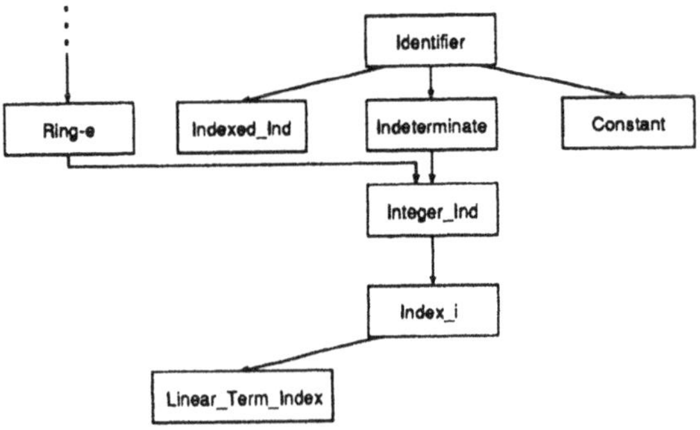

Figure 1: **Identifier** type and subtypes

The ADT **Index_i**, that comprises identifiers that might be accepted as indexes, is defined to be equal to the type **Integer_Ind** and, in particular, **Index_i** objects are ring elements.

Index_i =
    inherit   **Integer_Ind**;
    sorts   index_i, symbol, integer, integer_domain, boolean;
    operations
    from **Integer_Ind** redefining
        indeterminate_c as index_i_c: symbol integer → index_i
        name as index_i_name: index_i → symbol
        domain as index_i_domain: index_i → integer_domain
        value as index_i_value: index_i → integer
        equal? as index_i_equal?: index_i index_i → boolean
        depend_on_index as i_depend_on_index: index_i index_i → boolean
    proper
        change_index: index_i index_i index_i → index_i ;

The ADT **Index_i** is a supertype of the ADT of **Linear_Term_Index** that contains linear polynomials whose indeterminates belong to the type **Index_i** and range over the integers. They will represent the more general index object that we will consider. For instance, a **Linear_Term_Index** object is $k + h * 2$, where h, k are **Index_i** objects.

The specification for indexed variables is done according to the definition of *indexed object* given above. The ADT **Indexed_Ind(Domain)** specifies a recursive data type that redefines the name component, inherited by its supertype **Identifier**, to be of type **Indexed_Ind(Domain)**; and enriches its supertype by a domain, a value and a **Index_I** as index.

Indexed_Ind(Domain) =
    inherit   **Identifier**;

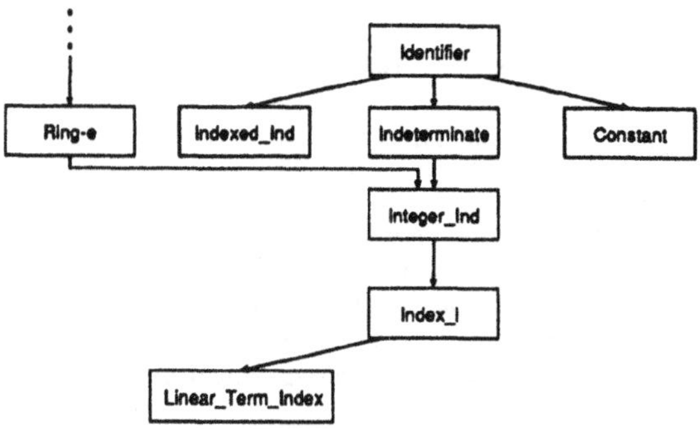

Figure 2: **Sigma** type and subtypes

<u>sorts</u> indexed_ind, index_i, symbol, domain, integer_domain, number, integer, boolean;

<u>operations</u>

from **Identifier** redefining

identifier_c as
indexed_ind_c: indexed_ind domain number index_i → indexed_ind
name as ind_i_name: indexed_ind → indexed_ind
equal? as ind_i_equal?: indexed_ind indexed_ind → boolean

proper

ind_i_domain: indexed_ind → domain
ind_i_value: indexed_ind → number
index_name: indexed_ind → symbol
index_domain: indexed_ind → integer_domain
index_value: indexed_ind → integer
index_simplify: indexed_ind → indexed_ind
change_index: indexed_ind index_i index_i → indexed_ind
depend_on_index: indexed_ind index_i → boolean;

The ADT **Indexed_Fun** is specified in a similar way given the specification of the ADT **Function** characterized by a name, a codomain and an argument.

The specification of the ADT **Sigma(Range)** describes the abstract type, related to the sigma notation defined above. It is characterized by a summand, that is a ring element, and a range. According to the different ways in which the range can be specified, the distinct subtypes associated to series or associated to sums are derived.

Sigma(Range)=

<u>sorts</u> range, ring-e, sigma , boolean, number;

<u>operations</u>

proper

sigma_c: ring-e range → sigma
summand: sigma → ring-e

index: sigma → index_i
sigma_range: sigma → range
axioms ∀s: sigma,
(sigma_c(summand(s), sigma_range(s)) = s);

The type **Sigma(Range)** is a supertype for two subtypes: the type **Series(Infinite_Range)** and the type **Sum(Finite_Range)**. The former is the ADT to which all manipulation laws can be applied as it inherits the properties of **Ring-e**. Some manipulation laws can also be defined for the type **Conv_Series(Infinite_Range)** as pointed out in Section 2.3, but we omit the specification here because it resembles that of **Sum(Finite_Range)**:

Sum(Finite_Range)=
inherit Ring_e, Sigma(Range);
sorts finite_range, ring-e, sum, boolean, number;
operations
from Ring-e redefining
0 as 0_s: → sum
1 as 1_s: → sum
− as $-_s$: sum sum → sum
+ as $+_s$: sum sum → sum
* as $*_s$: sum sum → sum;
from Sigma(Range) redefining
sigma_c as sum_c: ring-e finite_range → sum
summand as sum_summand: sum → ring-e
index as sum_index: sum → index_i
sigma_range as sum_range: sum → finite_range
proper
eval: sum → ring-e
distribute: sum sum → sum sum
interchange_order: sum sum → sum
manipulate_domain: sum sum → sum sum
change_index: sum index_i index_i → sum;
axioms ∀s: sum,
(sum_c(sum_summand(s), sum_range(s)) = s);

For instance, a finite sum whose range is conventionally a finite interval, is specified as follows:

Finite_Sum = **Sum**(Finite_Interval);

These specifications of ADT's, that are related to each other in a modular and in a supertype/subtype relation, can be directly implemented using an object-oriented environment. We have experimented this possibility adopting the Common Lisp substrate [Ste84] because it is equipped with a garbage collector, it is highly portable and it is easily extendible by symbolic manipulation facilities already available.

4 Experimenting with CLOS

The Common Lisp object-oriented extension, Common Lisp Object System, provides generic functions, multiple inheritance and method combination to deal with the fundamental objects represented by the classes. A class determines the structure and the set of operations of an object. A generic function is a function whose behavior depends on the classes of its arguments: this class-specific behavior is defined by the methods that contain the code to be executed when the dispacher has chosen and called the right one depending on the classes of the actual parameters of the generic function.

Classes in CLOS are objects that determine, by inheritance from other classes, structure and behavior of other run-time objects called instances. The specification given in the previous section can be directly translated to CLOS code by using classes to implement the abstract data types and methods to implement operations on them.

The macro defclass is used to set up a new class definition by giving the name of the class, the list of its superclasses, a set of slot specifiers and a set of class options. A class inherits all the slots and all the methods defined by its superclasses. As an example, the class meeting the specification of **Indexed_Ind** looks like follows.

```
(defclass indexed-ind (identifier)
  ((name  :initarg :name
          :accessor name
          :type indexed-ind)
   (domain :initarg :domain
           :accessor id-domain)
   (value :initarg :value
          :accessor id-value)
   (index :initarg :index
          :accessor index
          :type index-i)))
```

Objects can be created that belong to these classes and assigned to global variables in order to manipulate them:

```
> (show xi)
(X [ I ])
> (setq y xi)
#<Real-Indexed-Ind #XA2BB5B>

> (setq sum1 (fsum-c i13 y))
#<Finite-Sum #XA36AD3>
> (show sum1)
summand is: (X [ I ])
range is: ([ 1 <= I <= 3 ])
```

Methods implement operations according to the class of the arguments, for example, the operation that renames an index has different behavior when the renaming is done on a **Finite_Sigma** object or on an **Indexed_Ind** object and therefore two distinct versions of the operation `change-index` are supplied.

```
(defmethod change-index ((oldindex index-i)
                         (newindex index-i)
                         (fs finite-sum))
  (setf (sum-range fs)
        (change-index oldindex newindex (sum-range fs)))
  (cond ((depend-on-index  oldindex (summand fs))
         (setf (summand fs)
               (change-index oldindex newindex (summand fs)))))
  fs)

(defmethod change-index ((oldindex index-i)
                         (newindex index-i)
                         (indexed indexed-ind))
  (setf (index indexed)
        (change-index oldindex newindex (index indexed)))
  indexed)
```

Renaming the index of the object associated to the variable `sum1` performs as follows:

```
> (change-index sum1 i j)
#<Finite-Sum #XA36AD3>
> (show sum1)
summand is: (X [ J ])
range is: ([ 1 <= J <= 3 ])
```

Up to now, our experiments are limited to the implementation of manipulation rules that deal with the simpler case of a finite sum whose range is given through an interval. We have, therefore, also completed a package for the data type describing finite intervals and operations on them like intersection, union and splitting that we have not discussed here.

5 Concluding Remarks and Future Works

We have presented the methodological experiments done in studying indexed objects, sums and series as a proper example for testing the integrated system proposed in [LT91]. As previously remarked, our aim is to define a set of rules to manipulate sums in the sense of manipulating either the summand or the relation over the index. At present, the general rules implemented are limited to finite sums and could be extended also to infinite sums, provided that convergence is ensured.

The current computer-algebra systems are usually equipped with a library of special case series like power series representations of functions and, to our knowledge, no available system includes a mechanism for testing convergence. It can be discussed whether a check of convergence should be included in the package in order to safely manipulate series. This would mean including some well known criteria for the convergence of series, such as D'Alembert or Cauchy's, whose application can be directed by the user.

We have carried out our experiments using CLOS but we are aware that this object-oriented programming environment still lacks some capabilities which would be useful in implementing designed mathematical data types. In particular, dynamic typing is the missing feature which we had to simulate because it was needed in the instantiation of parametric classes and it would be of interest to extend CLOS in this direction.

6 Acknowledgments

I would like to thank Dongming Wang, Alfonso Miola and the research group in Rome for discussions, suggestions and pleasant working time we had.

References

[BDG+88] D. G. Bobrow, L. G. DeMichiel, R. P. Gabriel, K. Kahn, S. E. Keene, G. Kiczales, D. A. Moon, M. Stefik, and D. L. Weinreb. Common Lisp Object System Specification. Technical Report 88-002, X3J13, ANSI Common Lisp Standard Committee, July 1988.

[Cap91] O. Caprotti. A Formal Characterization of Indexed Mathematical Objects for the Manipulation of Sums and Series. Technical Report RISC-Linz Series no. 91-26.0, RISC-Linz, Research Institute for Symbolic Computation, Linz, Austria, June 1991.

[CM90] G. Cioni and A. Miola. Specification and Programming Methodologies for Axiomatizable Objects Manipulation: TASSO Project. Technical report, Istituto di Analisi dei Sistemi ed Informatica, CNR, Roma, Italy, 1990.

[JT81] R. D. Jenks and B. M. Trager. A Language for Computational Algebra. In *Symposium on Symbolic and Algebraic Manipulation, ACM*, Snowbird, Utah, August 1981.

[LT91] C. Limongelli and M. Temperini. Abstract specification of structures and methods in symbolic mathematical computation. *Theoretical Computer Science*, 1991. To appear.

[Mio83] A. Miola. Design Specifications for Manipulation of Sums and Series. Technical Report R.66, Istituto di Analisi dei Sistemi ed Informatica, CNR, Roma, Italy, 1983.

[Mio84] A. Miola. An Interactive System for Manipulation of Sums and Series. In *International Conference on Artificial Intelligence, AIMSA*, Varna, September 1984.

[Ste84] G. L. Jr. Steele. *Common Lisp, The Language.* Digital Press, Burlington, MA, 1984.

[SWW92] W. Shen, B. Wall, and D. Wang. Manipulating Uncertain Mathematical Objects: The Case of Indefinite Sums and Products. Proc. DISCO'92 (this volume), 1992.

[Wan90] D. Wang. Differentiation and Integration of Indefinite Summations. Technical Report RISC-Linz Series no. 90-37.0, RISC-Linz, Research Institute for Symbolic Computation, Linz, Austria, August 1990.

[Wan91] D. Wang. A Toolkit for Manipulating Indefinite Summations with Application to Neural Networks. *ACM SIGSAM Bulletin*, 25(3):18–27, 1991.

Author Index

Springer-Verlag and the Environment

We at Springer-Verlag firmly believe that an international science publisher has a special obligation to the environment, and our corporate policies consistently reflect this conviction.

We also expect our business partners – paper mills, printers, packaging manufacturers, etc. – to commit themselves to using environmentally friendly materials and production processes.

The paper in this book is made from low- or no-chlorine pulp and is acid free, in conformance with international standards for paper permanency.

GH 155 x 235 mm

Lecture Notes in Computer Science

For information about Vols. 1–650
please contact your bookseller or Springer-Verlag

Vol. 651: R. Koymans, Specifying Message Passing and Time-Critical Systems with Temporal Logic. IX, 164 pages. 1992.

Vol. 652: R. Shyamasundar (Ed.), Foundations of Software Technology and Theoretical Computer Science. Proceedings, 1992. XIII, 405 pages. 1992.

Vol. 653: A. Bensoussan, J.-P. Verjus (Eds.), Future Tendencies in Computer Science, Control and Applied Mathematics. Proceedings, 1992. XV, 371 pages. 1992.

Vol. 654: A. Nakamura, M. Nivat, A. Saoudi, P. S. P. Wang, K. Inoue (Eds.), Parallel Image Analysis. Proceedings, 1992. VIII, 312 pages. 1992.

Vol. 655: M. Bidoit, C. Choppy (Eds.), Recent Trends in Data Type Specification. X, 344 pages. 1993.

Vol. 656: M. Rusinowitch, J. L. Rémy (Eds.), Conditional Term Rewriting Systems. Proceedings, 1992. XI, 501 pages. 1993.

Vol. 657: E. W. Mayr (Ed.), Graph-Theoretic Concepts in Computer Science. Proceedings, 1992. VIII, 350 pages. 1993.

Vol. 658: R. A. Rueppel (Ed.), Advances in Cryptology – EUROCRYPT '92. Proceedings, 1992. X, 493 pages. 1993.

Vol. 659: G. Brewka, K. P. Jantke, P. H. Schmitt (Eds.), Nonmonotonic and Inductive Logic. Proceedings, 1991. VIII, 332 pages. 1993. (Subseries LNAI).

Vol. 660: E. Lamma, P. Mello (Eds.), Extensions of Logic Programming. Proceedings, 1992. VIII, 417 pages. 1993. (Subseries LNAI).

Vol. 661: S. J. Hanson, W. Remmele, R. L. Rivest (Eds.), Machine Learning: From Theory to Applications. VIII, 271 pages. 1993.

Vol. 662: M. Nitzberg, D. Mumford, T. Shiota, Filtering, Segmentation and Depth. VIII, 143 pages. 1993.

Vol. 663: G. v. Bochmann, D. K. Probst (Eds.), Computer Aided Verification. Proceedings, 1992. IX, 422 pages. 1993.

Vol. 664: M. Bezem, J. F. Groote (Eds.), Typed Lambda Calculi and Applications. Proceedings, 1993. VIII, 433 pages. 1993.

Vol. 665: P. Enjalbert, A. Finkel, K. W. Wagner (Eds.), STACS 93. Proceedings, 1993. XIV, 724 pages. 1993.

Vol. 666: J. W. de Bakker, W.-P. de Roever, G. Rozenberg (Eds.), Semantics: Foundations and Applications. Proceedings, 1992. VIII, 659 pages. 1993.

Vol. 667: P. B. Brazdil (Ed.), Machine Learning: ECML – 93. Proceedings, 1993. XII, 471 pages. 1993. (Subseries LNAI).

Vol. 668: M.-C. Gaudel, J.-P. Jouannaud (Eds.), TAPSOFT '93: Theory and Practice of Software Development. Proceedings, 1993. XII, 762 pages. 1993.

Vol. 669: R. S. Bird, C. C. Morgan, J. C. P. Woodcock (Eds.), Mathematics of Program Construction. Proceedings, 1992. VIII, 378 pages. 1993.

Vol. 670: J. C. P. Woodcock, P. G. Larsen (Eds.), FME '93: Industrial-Strength Formal Methods. Proceedings, 1993. XI, 689 pages. 1993.

Vol. 671: H. J. Ohlbach (Ed.), GWAI-92: Advances in Artificial Intelligence. Proceedings, 1992. XI, 397 pages. 1993. (Subseries LNAI).

Vol. 672: A. Barak, S. Guday, R. G. Wheeler, The MOSIX Distributed Operating System. X, 221 pages. 1993.

Vol. 673: G. Cohen, T. Mora, O. Moreno (Eds.), Applied Algebra, Algebraic Algorithms and Error-Correcting Codes. Proceedings, 1993. X, 355 pages 1993.

Vol. 674: G. Rozenberg (Ed.), Advances in Petri Nets 1993. VII, 457 pages. 1993.

Vol. 675: A. Mulkers, Live Data Structures in Logic Programs. VIII, 220 pages. 1993.

Vol. 676: Th. H. Reiss, Recognizing Planar Objects Using Invariant Image Features. X, 180 pages. 1993.

Vol. 677: H. Abdulrab, J.-P. Pécuchet (Eds.), Word Equations and Related Topics. Proceedings, 1991. VII, 214 pages. 1993.

Vol. 678: F. Meyer auf der Heide, B. Monien, A. L. Rosenberg (Eds.), Parallel Architectures and Their Efficient Use. Proceedings, 1992. XII, 227 pages. 1993.

Vol. 679: C. Fermüller, A. Leitsch, T. Tammet, N. Zamov, Resolution Methods for the Decision Problem. VIII, 205 pages. 1993. (Subseries LNAI).

Vol. 680: B. Hoffmann, B. Krieg-Brückner (Eds.), Program Development by Specification and Transformation. XV, 623 pages. 1993.

Vol. 681: H. Wansing, The Logic of Information Structures. IX, 163 pages. 1993. (Subseries LNAI).

Vol. 682: B. Bouchon-Meunier, L. Valverde, R. R. Yager (Eds.), IPMU '92 – Advanced Methods in Artificial Intelligence. Proceedings, 1992. IX, 367 pages. 1993.

Vol. 683: G.J. Milne, L. Pierre (Eds.), Correct Hardware Design and Verification Methods. Proceedings, 1993. VIII, 270 Pages. 1993.

Vol. 684: A. Apostolico, M. Crochemore, Z. Galil, U. Manber (Eds.), Combinatorial Pattern Matching. Proceedings, 1993. VIII, 265 pages. 1993.

Vol. 685: C. Rolland, F. Bodart, C. Cauvet (Eds.), Advanced Information Systems Engineering. Proceedings, 1993. XI, 650 pages. 1993.

Vol. 686: J. Mira, J. Cabestany, A. Prieto (Eds.), New Trends in Neural Computation. Proceedings, 1993. XVII, 746 pages. 1993.

Vol. 687: H. H. Barrett, A. F. Gmitro (Eds.), Information Processing in Medical Imaging. Proceedings, 1993. XVI, 567 pages. 1993.

Vol. 688: M. Gauthier (Ed.), Ada-Europe '93. Proceedings, 1993. VIII, 353 pages. 1993.

Vol. 689: J. Komorowski, Z. W. Ras (Eds.), Methodologies for Intelligent Systems. Proceedings, 1993. XI, 653 pages. 1993. (Subseries LNAI).

Vol. 690: C. Kirchner (Ed.), Rewriting Techniques and Applications. Proceedings, 1993. XI, 488 pages. 1993.

Vol. 691: M. Ajmone Marsan (Ed.), Application and Theory of Petri Nets 1993. Proceedings, 1993. IX, 591 pages. 1993.

Vol. 692: D. Abel, B.C. Ooi (Eds.), Advances in Spatial Databases. Proceedings, 1993. XIII, 529 pages. 1993.

Vol. 693: P. E. Lauer (Ed.), Functional Programming, Concurrency, Simulation and Automated Reasoning. Proceedings, 1991/1992. XI, 398 pages. 1993.

Vol. 694: A. Bode, M. Reeve, G. Wolf (Eds.), PARLE '93. Parallel Architectures and Languages Europe. Proceedings, 1993. XVII, 770 pages. 1993.

Vol. 695: E. P. Klement, W. Slany (Eds.), Fuzzy Logic in Artificial Intelligence. Proceedings, 1993. VIII, 192 pages. 1993. (Subseries LNAI).

Vol. 696: M. Worboys, A. F. Grundy (Eds.), Advances in Databases. Proceedings, 1993. X, 276 pages. 1993.

Vol. 697: C. Courcoubetis (Ed.), Computer Aided Verification. Proceedings, 1993. IX, 504 pages. 1993.

Vol. 698: A. Voronkov (Ed.), Logic Programming and Automated Reasoning. Proceedings, 1993. XIII, 386 pages. 1993. (Subseries LNAI).

Vol. 699: G. W. Mineau, B. Moulin, J. F. Sowa (Eds.), Conceptual Graphs for Knowledge Representation. Proceedings, 1993. IX, 451 pages. 1993. (Subseries LNAI).

Vol. 700: A. Lingas, R. Karlsson, S. Carlsson (Eds.), Automata, Languages and Programming. Proceedings, 1993. XII, 697 pages. 1993.

Vol. 701: P. Atzeni (Ed.), LOGIDATA+: Deductive Databases with Complex Objects. VIII, 273 pages. 1993.

Vol. 702: E. Börger, G. Jäger, H. Kleine Büning, S. Martini, M. M. Richter (Eds.), Computer Science Logic. Proceedings, 1992. VIII, 439 pages. 1993.

Vol. 703: M. de Berg, Ray Shooting, Depth Orders and Hidden Surface Removal. X, 201 pages. 1993.

Vol. 704: F. N. Paulisch, The Design of an Extendible Graph Editor. XV, 184 pages. 1993.

Vol. 705: H. Grünbacher, R. W. Hartenstein (Eds.), Field-Programmable Gate Arrays. Proceedings, 1992. VIII, 218 pages. 1993.

Vol. 706: H. D. Rombach, V. R. Basili, R. W. Selby (Eds.), Experimental Software Engineering Issues. Proceedings, 1992. XVIII, 261 pages. 1993.

Vol. 707: O. M. Nierstrasz (Ed.), ECOOP '93 – Object-Oriented Programming. Proceedings, 1993. XI, 531 pages. 1993.

Vol. 708: C. Laugier (Ed.), Geometric Reasoning for Perception and Action. Proceedings, 1991. VIII, 281 pages. 1993.

Vol. 709: F. Dehne, J.-R. Sack, N. Santoro, S. Whitesides (Eds.), Algorithms and Data Structures. Proceedings, 1993. XII, 634 pages. 1993.

Vol. 710: Z. Ésik (Ed.), Fundamentals of Computation Theory. Proceedings, 1993. IX, 471 pages. 1993.

Vol. 711: A. M. Borzyszkowski, S. Sokołowski (Eds.), Mathematical Foundations of Computer Science 1993. Proceedings, 1993. XIII, 782 pages. 1993.

Vol. 712: P. V. Rangan (Ed.), Network and Operating System Support for Digital Audio and Video. Proceedings, 1992. X, 416 pages. 1993.

Vol. 713: G. Gottlob, A. Leitsch, D. Mundici (Eds.), Computational Logic and Proof Theory. Proceedings, 1993. XI, 348 pages. 1993.

Vol. 714: M. Bruynooghe, J. Penjam (Eds.), Programming Language Implementation and Logic Programming. Proceedings, 1993. XI, 421 pages. 1993.

Vol. 715: E. Best (Ed.), CONCUR'93. Proceedings, 1993. IX, 541 pages. 1993.

Vol. 716: A. U. Frank, I. Campari (Eds.), Spatial Information Theory. Proceedings, 1993. XI, 478 pages. 1993.

Vol. 717: I. Sommerville, M. Paul (Eds.), Software Engineering – ESEC '93. Proceedings, 1993. XII, 516 pages. 1993.

Vol. 718: J. Seberry, Y. Zheng (Eds.), Advances in Cryptology – AUSCRYPT '92. Proceedings, 1992. XIII, 543 pages. 1993.

Vol. 719: D. Chetverikov, W.G. Kropatsch (Eds.), Computer Analysis of Images and Patterns. Proceedings, 1993. XVI, 857 pages. 1993.

Vol. 720: V.Mařík, J. Lažanský, R.R. Wagner (Eds.), Database and Expert Systems Applications. Proceedings, 1993. XV, 768 pages. 1993.

Vol. 721: J. Fitch (Ed.), Design and Implementation of Symbolic Computation Systems. Proceedings, 1992. VIII, 215 pages. 1993.

Vol. 722: A. Miola (Ed.), Design and Implementation of Symbolic Computation Systems. Proceedings, 1993. XII, 384 pages. 1993.

Vol. 723: N. Aussenac, G. Boy, B. Gaines, M. Linster, J.-G. Ganascia, Y. Kodratoff (Eds.), Knowledge Acquisition for Knowledge-Based Systems. Proceedings, 1993. XIII, 446 pages. 1993. (Subseries LNAI).

Vol. 724: P. Cousot, M. Falaschi, G. Filè, A. Rauzy (Eds.), Static Analysis. Proceedings, 1993. IX, 283 pages. 1993.

Vol. 725: A. Schiper (Ed.), Distributed Algorithms. Proceedings, 1993. VIII, 325 pages. 1993.

Vol. 726: T. Lengauer (Ed.), Algorithms — ESA '93. Proceedings, 1993. IX, 419 pages. 1993.